FOCUS ON

Grades 5-8

MIDDLE SCHOOL

Biology

3rd Edition

Rebecca W. Keller, PhD

Real Science-4-Kids

Cover design: David Keller
Opening page: David Keller
Text illustrations: Janet Moneymaker, Rebecca W. Keller, PhD

Focus On Middle School Biology Student Textbook–3rd Edition (hardcover)
ISBN 978-1-941181-69-0

Published by Gravitas Publications Inc.
www.gravitaspublications.com
www.realscience4kids.com

Contents

Chapter 1 What Is Biology?

1.1 Introduction

Biology is the study of life. The word biology comes from the Greek words *bios*, which means "life," and *logos*, which means "description." Biology is the field of science that "describes life." Biology is concerned with all living things and how they interact with one another.

Living creatures come in many different sizes, shapes, and colors. Some are big and some are very small. Some are green, some are red, some are black, and some are white. Some see with two eyes, some see with eight eyes, and some have no eyes at all! Some fly, some walk, some swim, and some crawl.

There are many different kinds of living things, but they all have one thing in common. They all are alive. But what does it mean for something to be alive?

Both living things and nonliving things are made of the same material—atoms! But if living things and nonliving things are all made of atoms, why are they so different? Why can a butterfly land on a rock, but a rock cannot fly away to find food?

1.2 What Is Life?

It seems that defining life should be easy. Even a young child knows the difference between living things and nonliving things. But finding a definition for life is actually very difficult!

One way to define life is to list the properties that are unique to living things. For example, living things have the ability to grow, the ability to reproduce, and the ability to adapt to the environment. However, a computer program can be designed to grow, reproduce, and adapt to the environment, yet we wouldn't say a computer program is alive. There must be something missing from this list of properties. So what else is needed to define life?

ARISTOTLE
384-322 B.C.E.

HIPPOCRATES
circa 460-circa 377 B.C.E.

The struggle to define life goes back many centuries. The Greek philosophers thought a lot about life and how to define life. Aristotle, a Greek philosopher (384-322 B.C.E.), believed that living things have a moving principle, which he defined as a force that causes an object to become itself.

Galen (circa 129-circa 199 C.E.), a Greek physician who studied anatomy, agreed with Aristotle and further developed the idea of life having a moving principle. He referred to the moving principle as the vital spirit.

GALEN
circa 129-circa 199 C.E.

Not all of the Greek philosophers agreed with the idea of a moving principle or a vital spirit. Hippocrates (circa 460-circa 377 B.C.E.), an early Greek physician, disagreed with Aristotle. He said that life is not caused by a moving principle, but by the ether, which he said was a type of fire that always existed and is present in air and in other matter.

DEMOCRITUS
Circa 460-370 B.C.E.

Other Greek philosophers, called atomists, believed that life is simply the result of movements and combinations of small invisible, indestructible particles. Probably the most famous atomist was Democritus (circa 460-370 B.C.E.). Democritus proposed that all matter is composed of indivisible particles called atoms.

Many new ideas about life developed in the 16th and 17th centuries. These ideas usually combined some sort of mechanical theory (the idea that living things function like machines) with some explanation of purpose (why the living things exist).

Rene Descartes (1596-1650 C.E.) was a French philosopher who thought about how atoms form molecules. He also developed a "mechanical philosophy," or the idea of mechanism. He believed that all living creatures are like machines and their behaviors are controlled solely by forces pushing the organs of the body.

DESCARTES
1596-1650 C.E.

During the 1800s as we learned much more about the cell, it became much easier to explain life in terms of chemical reactions. By the end of the 20th century, a completely non-vitalist philosophy had emerged. This idea is called materialism. According to materialism, everything is made of matter only, so all life can be explained solely by the laws of chemistry and physics.

A particular facet of materialism is called reductionism. Reductionism is the belief that because life can be explained by the laws of chemistry and physics (materialism) you can completely understand something by studying its parts. For example, if you don't know what a bicycle is, then you can take it apart, and by understanding the tires, the spokes, and the gears, you can understand a bicycle.

Reductionism and materialism have played a very important role in shaping scientific understanding, but not everything can be explained by looking only at the individual parts. Systems biology, for example, involves examining the interactive system, not only the parts that make up the system. As more information about life is discovered, new ways of thinking about and studying life are sure to emerge.

1.3 Philosophical Maps Help Us Interpret Science

All of these "-isms" are particular ways to interpret the world. Vitalism, materialism, mechanism, and reductionism are philosophical maps that help us get a clearer picture of the world around us. Just like physical maps help us navigate directions in cities, philosophical maps help us interpret and understand scientific data.

However, it's important that we don't confuse the map with reality. A map is just a map, and although it is useful, it does not always give the most accurate picture of reality. Also, the best way to navigate any territory is to use more than one map.

Often scientists disagree about which map is "right" for understanding science. However, there is not one "right" map for all questions. Materialism and reductionism can be useful in answering some questions, and vitalism and mechanism can be useful for answering other questions. The scientist who understands and can use multiple maps has a better chance of seeing the world more clearly than the scientist who only uses one map. Also, the scientist who uses many maps will find more opportunities to make new discoveries.

1.4 Organizing Life

Taxonomy

One way to understand living things is to organize or classify them. By organizing the different types of living things into groups, scientists can better study both their similarities and their differences.

The branch of biology concerned with naming and classifying the many different types of living things is called taxonomy. Carolus Linnaeus (1707-1778 C.E.), a Swedish physician, was the founder of taxonomy. Linnaeus viewed science as a way to understand how the world is organized. He began to carefully study all the living things he could find. Whenever he found animals that were similar, like dogs and wolves or bees and wasps, he grouped them together. Grouping things together is what is meant by classifying. A new creature is classified in a group depending on which creatures it has the most in common with. Sometimes it is very hard to decide which group a creature fits into.

LINNAEUS
1707-1778 C.E.

Domains and Kingdoms

Because there are so many different kinds of living creatures, it has been hard for scientists to figure out exactly how to organize them. Several different approaches are currently in use. Until recently, the most commonly used approach divided all living things into five kingdoms.

However, modern taxonomy is beginning to use a system introduced in 1990 by Carl Woese. In this system, living things are divided into three domains which are then further divided into six kingdoms. The three domains are called Eukarya, Bacteria, and Archaea.

The kingdoms in those domains are Protista, Plantae, Fungi, Animalia, Bacteria (also called Eubacteria), and Archaea (also called Archaebacteria).

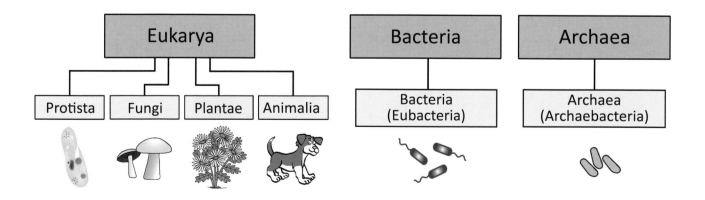

Taxonomy is continuing to change as scientists make new discoveries about living things, and scientists may use different taxonomic systems according to what they are trying to find out about living things.

How do we decide in which domain and which kingdom a living thing should be placed?

Should a dog be grouped with the elephants, or should it be placed with the bacteria?

Should a house cat be grouped with the house plants or with the bunnies?

What about a lizard? Is it like a mushroom or like a jellyfish?

Before placing a living thing into a particular kingdom, it must first be placed in a domain. It is primarily the difference in the structure of the cells that ultimately determines the domain in which an organism is placed.

Dog cells are more like elephant cells than they are like bacteria, so dogs are grouped with elephants in the domain Eukarya. Cat cells are more like bunny cells than archaeal cells, so cats are grouped with bunnies in the domain Eukarya. Lizards and jellyfish, although very different from each other, have similar cells, so lizards are grouped with jellyfish in the domain Eukarya and not grouped with bacteria or archaea.

Once an organism is placed into a domain, it is further categorized and placed in a kingdom. The animal kingdom, Animalia, includes ALL of the animals: dogs, cats, frogs, sea urchins, bees, birds, snakes, jellyfish, bunnies, and even us!

The animal kingdom has a wide variety of living creatures in it. Some are similar to each other, like dogs and wolves, and some are not so similar, like bees and snails, but ALL animals in the kingdom Animalia have animal cells. (See Chapter 5.) This distinguishes them from other living things.

The plant kingdom, Plantae, includes all plants: trees, grass, flowers, ferns, dandelions, and even asparagus! Again, some plants are similar to each other and some plants are very, very different, but ALL plants have plant cells. (See Chapter 5.)

The fungus kingdom, Fungi, includes mushrooms, toadstools, truffles, and even athlete's foot! The fungi were once grouped with plants, but they have many unique features and are now placed in a kingdom of their own.

The last three kingdoms, Protista, Bacteria (Eubacteria), and Archaea (Archaebacteria), include most of the microscopic organisms, such as paramecia and amoebas. These organisms cannot be seen with the unaided eye and were unknown before microscopes were invented.

The kingdom Protista is in the domain Eukarya because protists have cells similar to other Eukarya. In the kingdom Protista, there are

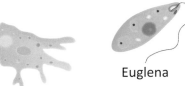

Amoeba

Euglena

Paramecium

Protista
Single and multicellular organisms like algae and paramecia

creatures that have both plant-like and animal-like features. Some, like euglena, are green and can use the Sun's energy to make food, like plants do. Others, like amoebas, catch and eat prey like animals do.

Bacteria and Archaea have cells that are different from each other and also from Eukarya; therefore, they have their own domains. Most of the organisms in the kingdoms Bacteria (Eubacteria) and Archaea (Archaebacteria) are unicellular. That is, they have only one cell. These organisms have a variety of shapes. The three most common shapes are spheres, rods, and spirals.

Bacteria and Archaea
Single-celled organisms

Rods

Spheres

Spirals

Further Classification

Once a living thing has been placed in a kingdom, the classification continues. Living things are further organized by being placed in additional categories that depend on a variety of criteria, like whether or not they have a backbone or whether or not they lay eggs. For example, although all animals are in the kingdom Animalia, it seems obvious that dogs and bees and snakes should be in different groups.

Kingdoms are divided into smaller groups called phyla. Dogs, frogs, and cats are members of the phylum Chordata because they all have backbones, and bees are in the phylum Arthropoda because they have "jointed feet (legs)."

In the same way, the phyla are divided into smaller groups called classes. Dogs and cats are all in the class Mammalia because they nurse their young, and frogs are in the class Amphibia because they live both in water and on land.

Classes are further divided into orders. Both cats and dogs are in the order Carnivora because they eat meat. Sometimes orders are divided into suborders. The order Carnivora is divided into the suborder Feliformia for cat-like animals and the suborder Caniformia for dog-like animals. The suborders are then divided into families. Cats are in the family Felidae, and dogs are in the family Canidae.

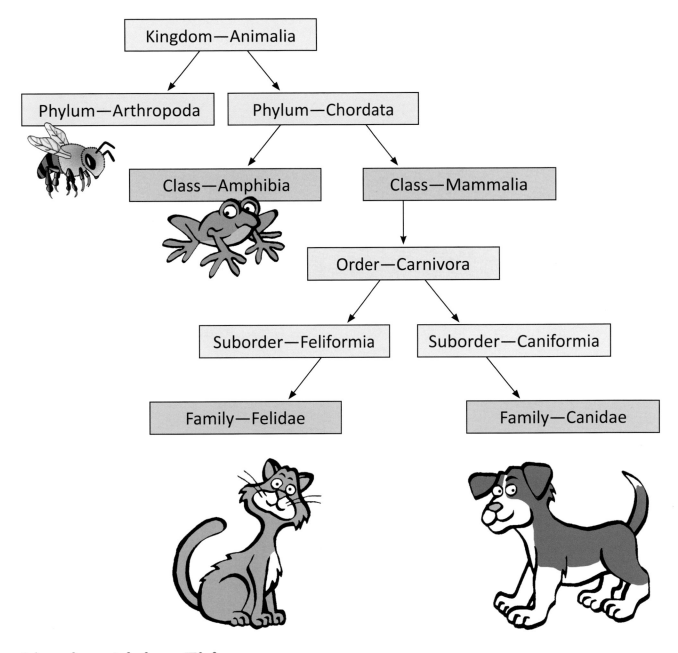

Naming Living Things

Finally, families are further divided into the genus, and the genus is divided into the species. The genus is the last group in which a living creature is placed, and the species identifies each creature placed in the genus, so each different type of living thing has a unique genus and species name. For example, both a bobcat and a house cat are in the genus *Felis*. A bobcat has the species name *rufa,* and a house cat has the species name *catus.* So a house cat is a *Felis catus* and a bobcat is a *Felis rufa*.

A tiger is a kind of cat, but it is different from both bobcats and house cats. It is in the genus *Panthera* and has a species name *tigris*. So, a tiger is called a *Panthera tigris*. A lion is like a tiger and is also in the genus *Panthera,* but it has a species name *leo*, so a lion is a *Panthera leo*.

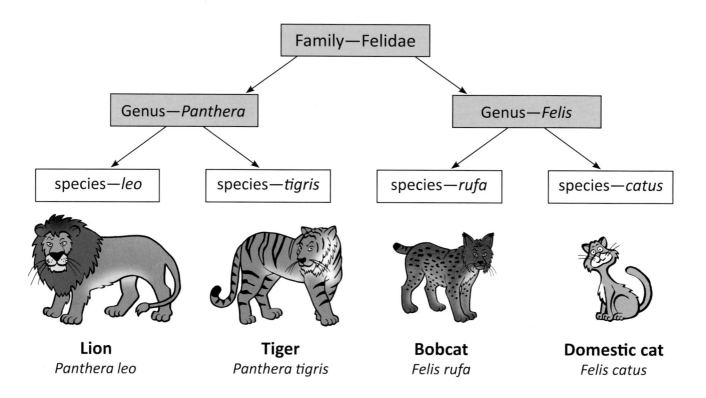

Lion
Panthera leo

Tiger
Panthera tigris

Bobcat
Felis rufa

Domestic cat
Felis catus

All living things have a particular genus and species name. The name for household dogs is *Canis familiaris,* and for humans it is *Homo sapiens* from the Latin words meaning "man wise." Note that the genus and species are written in italics, and the genus is capitalized.

1.5 Summary

- Providing an exact definition of life is difficult, and both scientists and philosophers have contributed.

- Greek philosophers such as Aristotle, Galen, Hippocrates, and Democritus had different ideas about what causes living things to be alive.

- Vitalism, materialism, mechanism, and reductionism are philosophical maps that help us explore the world around us.

- Taxonomy is the branch of biology that classifies living things.

- Living things are grouped into categories so scientists can learn more about how they are the same and how they are different. Also, if a new creature is discovered, for instance, on the deep ocean floor, placing it into a group of known creatures will help scientists better understand how it lives.

- Living things are placed in a group depending on many characteristics, including what kind of cells they have, whether they have hair or scales, and whether or not they lay eggs.

- Several different systems of taxonomy are in use today, and taxonomy continues to change as new discoveries are made.

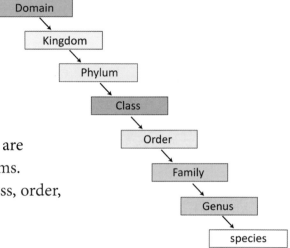

- All living things are classified into different groups. The largest group is the domain. There are three domains that are divided into six kingdoms. Kingdoms are further divided into phylum, class, order, family, genus, and species.

1.6 Some Things to Think About

- Make two lists, one of living things and one of nonliving things. Review your lists. What is different about the things that are living and those that are not alive?

- Review the different ways of defining life that are presented in this section. Do you think the way we define life will continue to change as new discoveries are made? Why or why not?

- How would you define life?

- What are some philosophical maps that you think you have used? Do you think these philosophical maps are helpful to you? Why or why not?

- Which kingdom would you most like to study? Why do you think this kingdom would be the most interesting one?

Chapter 2 Technology in Biology

A field biologist studying birds
Photo Credit: Travis Booms/USFWS

2.1 Introduction

The study of plants, animals, bacteria and viruses has come a long way from the days of Aristotle. Modern biology is a broad and diverse field of inquiry that includes botany (the study of plants), zoology (the study of animals), cell biology (the study of cells), molecular biology and genetics (the study of biological molecules and DNA), anatomy and physiology (the

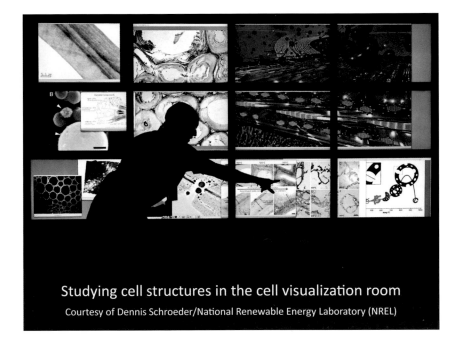

Studying cell structures in the cell visualization room
Courtesy of Dennis Schroeder/National Renewable Energy Laboratory (NREL)

study of animal and plant structure and functions), and many subcategories that include immunology (the study of the immune system), protistology (the study of protists), and marine biology (the study of marine animals and plants), to name a few.

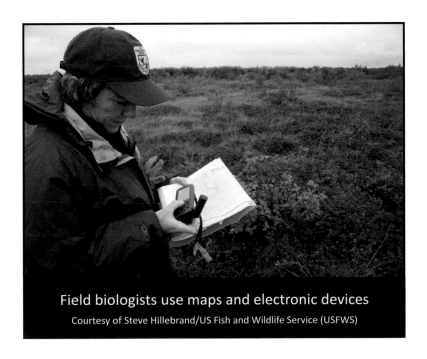

Field biologists use maps and electronic devices
Courtesy of Steve Hillebrand/US Fish and Wildlife Service (USFWS)

Because biology includes a large collection of specialized categories and subcategories, the technology biologists use can vary greatly from lab to lab. Some biology laboratories may use specialized equipment such as a light microscope or an electron microscope to investigate microscopic cellular structures and molecules. Other laboratories may exist entirely in the field where biologists study large animals, plants, or ecosystems. Some biology labs may look more like chemistry labs, complete with glass beakers, Erlenmeyer flasks, and graduated cylinders. Still other labs may look like military ships with radar equipment, tracking devices, and small submersibles for exploring the deep ocean.

Different types of microscopes are used in many biology labs, and we'll learn more about some of these microscopes in Chapter 3. However, because there is not a typical type of biology lab and because not all biology labs share similar equipment, we'll take a look at just a few different biology labs and the types of instruments they use.

A biologist tracks a black bear
Credit: John & Karen Hollingworth/USFWS

2.2 The Botany Laboratory

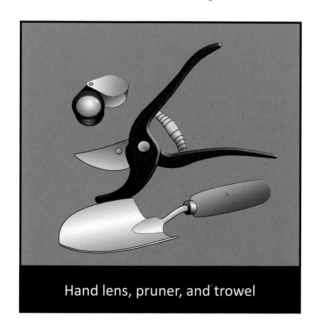

Hand lens, pruner, and trowel

Botany is the study of plants. Biologists who study plants are called botanists. Botanists who do field work may carry a backpack with a variety of different handheld tools such as a trowel for digging into the soil, a pruner for clipping woody samples, a hand lens for examining the small details of plants, a field notebook for recording observations, and a compass or GPS for finding the way back to civilization!

Once the botanist has collected samples, the next step might be to take them to a laboratory and use a microscope to examine the fine details of a flower or the cellular structure of the stem or leaf. Dissecting needles may be used to cut leaves, stems, or seeds so their internal structure can be observed. Once the botanist has observed the details of the flower, plant, or leaf, a plant press can be used to preserve the sample for future study.

Floristics is an area of study in botany in which a scientist observes the plants that live in a particular area, noting their number, types, relationships to each other, and how they are distributed throughout the area.

Floristic keys are reference tools used by researchers for identifying plants and plant parts and where the plants live. Floristic keys contain information that may include illustrations of the details of plants and their structure, lists of characteristics used to help identify the plant, seed, or flower being studied, and maps showing where a type of plant grows.

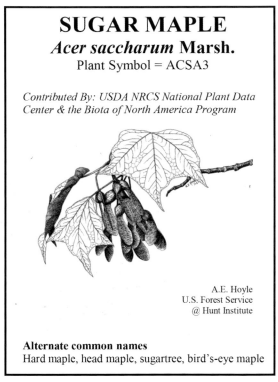

SUGAR MAPLE
Acer saccharum **Marsh.**
Plant Symbol = ACSA3

Contributed By: USDA NRCS National Plant Data Center & the Biota of North America Program

A.E. Hoyle
U.S. Forest Service
@ Hunt Institute

Alternate common names
Hard maple, head maple, sugartree, bird's-eye maple

2.3 The Molecular Biology and Genetics Laboratory

Molecular biology and genetics are the study of small molecules and molecular machines inside cells. In addition to the balances, graduated cylinders, and flasks found in many chemistry laboratories, molecular biologists and geneticists often use very specialized equipment. Because many processes inside cells cannot be observed directly with a microscope, molecular biologists and geneticists need to use methods and tools that will allow them to observe what happens inside cells without actually seeing the molecules in action.

Most molecular biology and genetics laboratories use small organisms like bacteria and yeast to study molecules such as proteins, DNA, and RNA, which are the main molecules involved in cell growth, reproduction, and cell death. Many laboratories grow bacterial cultures like *Escherichia coli (E. coli)* to see how they divide and change with the environment. To grow *E. coli*, a laboratory may have an autoclave to sterilize equipment, agar solutions and agar plates for growing and testing *E. coli* cultures, and large warm ovens for maintaining correct growth temperatures.

Proteins, DNA, and RNA can be removed from large volumes of bacterial solutions and analyzed using a method called gel electrophoresis. Gel electrophoresis is a process in which small amounts of proteins, DNA, or RNA are added to a thin gel placed between two glass or plastic plates. Because different sizes of proteins, DNA, and

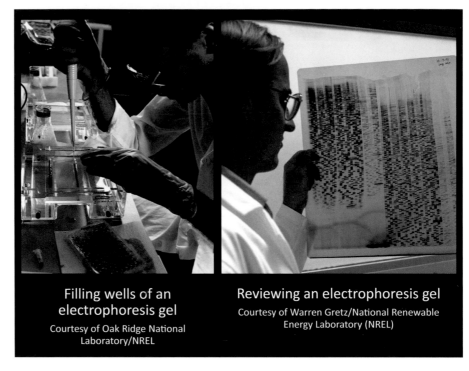

Filling wells of an electrophoresis gel
Courtesy of Oak Ridge National Laboratory/NREL

Reviewing an electrophoresis gel
Courtesy of Warren Gretz/National Renewable Energy Laboratory (NREL)

RNA will migrate through the gel at different speeds when an electrical current is passed through the gel, they will separate from one another and can be analyzed.

Gel Electrophoresis

1. A sample of DNA is cut into smaller fragments.

2. The DNA fragments are loaded into the wells of a gel that is held between two sheets of glass or plastic.

3. When an electric current is passed through the gel, the DNA fragments move down the gel according to size, with the largest fragments traveling the farthest.

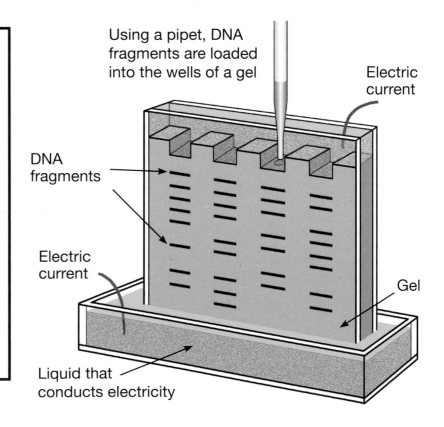

Using a pipet, DNA fragments are loaded into the wells of a gel

Electric current

DNA fragments

Electric current

Gel

Liquid that conducts electricity

2.4 The Marine Biology Laboratory

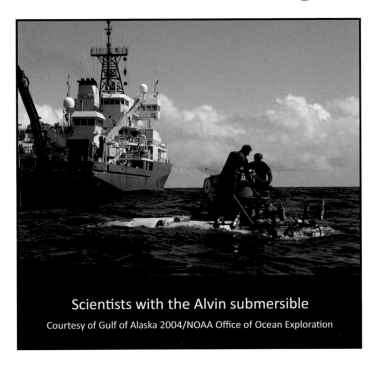

Scientists with the Alvin submersible

Courtesy of Gulf of Alaska 2004/NOAA Office of Ocean Exploration

Marine biology is the study of plants and animals in oceans and other saltwater environments. Marine biologists have floating laboratories on boats or other sea vessels from which they can collect samples and observe sea life. Once back on land, a marine biologist can further analyze the samples in a brick and mortar laboratory.

From a boat, marine biologists may collect samples with the use of a variety of tools and instruments, including water samplers, plankton nets, and sediment corers. With these tools marine biologists can compare water samples, observe the health of small organisms like plankton, test for pollution or other toxins, and take samples of the sediments on the ocean floor along with the organisms that live in the sediments.

To observe and collect samples below the surface of the ocean, marine biologists can use scuba equipment, diving bells, submersibles, and remotely operated vehicles (ROVs). Scuba equipment allows researchers to swim below the surface for short periods of time. With a diving bell, researchers can travel along the bottom of the ocean for extended periods of time. To go into the deep ocean and travel longer distances, manned submersibles or remotely operated vehicles can be used. By going deeper into the ocean, marine biologists can observe how organisms live in colder and darker water. Scientists believe that the majority of ocean life is still undiscovered.

Divers record algae diversity

Courtesy of Dr. Jean Kenyon, NOAA/ NMFS/PISC/CRED

Marine biologists can also observe the behavior of marine animals by using trackers that send data to satellites. By placing a tracker on a shark, dolphin, or whale, marine biologists can map how the animals move, eat, and reproduce. This is especially helpful for studying sharks to find out when they might be wandering close to a shore full of beachgoers!

Once marine biologists have collected water, plankton, sediment, or other samples from the ocean, they may travel to a land based lab to analyze them. Their lab may have equipment similar to a chemistry or biology lab with balances, graduated cylinders, flasks, and possibly a gel electrophoresis apparatus.

Collecting botany samples
Courtesy of USFWS

Scientists work with ocean sediment samples
Courtesy of Jeremy Potter/Hidden Ocean 2005 Expedition:
NOAA Office of Ocean Exploration

A scientist looks out the window of the Delta submersible
Courtesy of Robert Schwemmer, CINMS, NOAA

2.5 Summary

- Biology is a diverse field of study and includes many different subcategories.

- Because biology includes so many specialized categories and subcategories, the technology biologists use can vary greatly from lab to lab.

- Botanists study plants and may use microscopes, dissecting needles, and plant presses.

- Molecular biologists and geneticists use specialized equipment for processes such as gel electrophoresis to study DNA, RNA, and proteins.

- Marine biologists may have a laboratory on a boat and may use scuba gear, diving bells, submersibles, or remotely operated vehicles to collect samples.

2.6 Some Things to Think About

- How do you think the use of computers has helped biologists?
  Do you think all types of biology labs use computers? Why or why not?

- If you were a botanist, how would you use a floristic key, a pruner, and a hand lens together to learn about a plant you want to identify?

- How do you think electrophoresis helps biologists study DNA?

- If you were a marine biologist, which of the below do you think you would like best? Why?

  Scuba diving to study marine plants.

  Going in a submersible to study marine animal life.

  Working in a lab on a ship.

  Studying sediments from the ocean floor.

  Analyzing water samples to see what they contain.

  Tracking sharks.

  Observing how ocean currents affect marine life.

  Working in a lab on land to study samples from the sea.

  All of the above!

Chapter 3   The Microscope

3.1 Introduction

The microscope is an instrument that makes small objects appear bigger. From examining the smallest organisms to mapping the blueprint of the cell, the microscope has dramatically changed the way we understand living things.

The invention of the first microscope is generally credited to Zacharias Janssen, although like many inventions the origin of the microscope is often debated. Zacharias Janssen was born about 1580 C.E. in Middelburg, the Netherlands. He became a spectacle-maker and developed an expertise in shaping and forming glass lenses. It was widely known at the time that a single, curved glass lens could correct vision and magnify objects. However, a single lens

is limited as to how much it can magnify. In the late 1500s Zacharias and his father Hans Janssen solved this problem by experimenting with combining lenses. They discovered that by using two lenses together the magnification was greatly increased.

The instrument they created resembled a spyglass with two lenses housed in a cylindrical tube. Their instrument became the first primitive compound microscope. It is this advance in technology that eventually led not only to the modern microscope but also the discovery of a fascinating new world of microscopic biology.

About a century later, a Dutch lens maker named Anton van Leeuwenhoek (1632–1723 C.E.) perfected the polishing and grinding of lenses. Although he probably did not combine lenses to make a compound microscope, one of his lenses could magnify samples up to

Previous page: Scanning electron microscope image credits: 1. Bug eye, CDC/Janice Carr; 2. Exoskeleton surface of a mite, CDC/ William L. Nicholson, PhD, Cal Welbourn, PhD, Gary R. Mullen; 3. Mosquito Head, CDC/Paul Howell

300 times. Leeuwenhoek was the first person to observe bacteria, red blood cells, and tiny creatures in pond water. Around the same time, Robert Hooke (1635-1703 C.E.), an English scientist, improved upon the Janssen microscope and was able to see the outline of cells in thinly sliced pieces of cork. Hooke became well known for his book *Micrographia* in which he wrote about and illustrated his observations.

Today, there are several different kinds of microscopes. The kind of microscope the Janssens invented and Hooke improved on is called a light microscope. A light microscope uses light and the interaction of light with glass lenses to focus and magnify an object. In addition to the light microscope, scientists can now use electrons and probes to image small objects. The electron microscope uses a beam of electrons to magnify objects, and a class of microscopes called probe microscopes use a stylus,

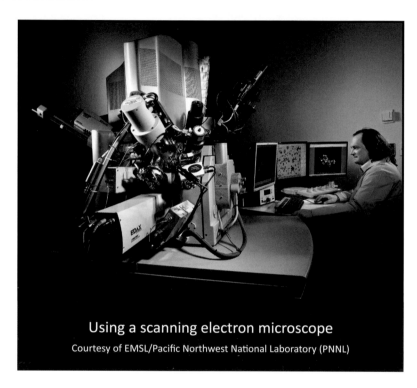

Using a scanning electron microscope
Courtesy of EMSL/Pacific Northwest National Laboratory (PNNL)

or small probe, to "feel" the features on the surface of a sample. We will learn more about light, electron, and probe microscopes later in this chapter.

3.2 The Size of Things

How big is an insect eye, a piece of hair, a protist, or a red blood cell? We can see a fly sitting on a wall, but with our naked eyes can we see mitochondria or a cell from the inside of our cheek? What size are the smallest things the human eye can see, and what size are the smallest things microscopes can help us see?

Before we take a look at different microscopes, it's important to explore the sizes of objects and the microscopic scale. The smallest object the unaided human eye can resolve is about the size of a strand of hair. Anything much smaller than a strand of hair needs to be magnified to be seen.

A Note About Units

The metric scale is the preferred system for measuring the size of objects with a microscope. Although metric units can be converted to British units, the British scale is never used as a measurement for microscopic objects. In the metric scale the meter is the base unit used to measure length. Units larger or smaller than the meter are multiples of ten and are given a prefix attached to the word *meter* to identify them. For example, a length 100 times smaller than the meter is called the centimeter. A length 1000 times smaller than the meter is called a millimeter. The following table lists the metric units most commonly used in microscopy.

Metric Measurement	Number of Times Smaller than a Meter	Measurement in Meters
1 centimeter (cm)	10^{-2} = 1 hundred times	0.01 meter
1 millimeter (mm)	10^{-3} = 1 thousand times	0.001 meter
1 micrometer (um)	10^{-6} = 1 million times	0.000001 meter
1 nanometer (nm)	10^{-9} = 1 billion times	0.000000001 meter

Insect Wing	Strand of Hair	Protist	Plant Cell	Animal Cell	Red Blood Cell	Bacterium	Mitochon-drion	Lysosome
6 mm 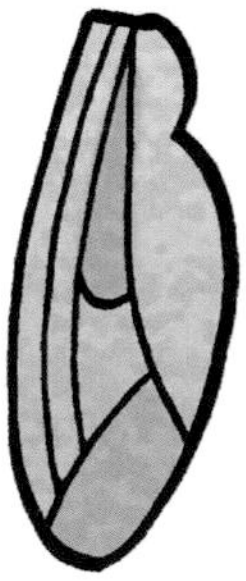	400 um 	200 um 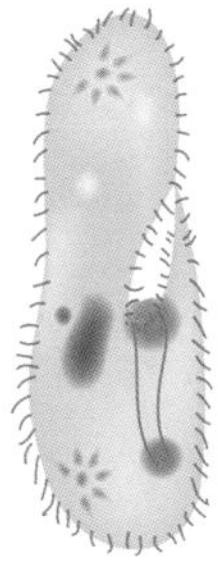	100 um 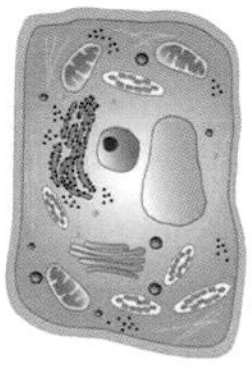	20 um 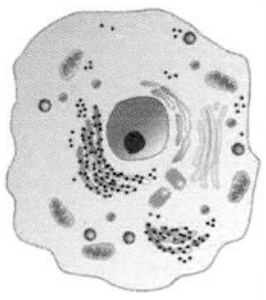	8 um 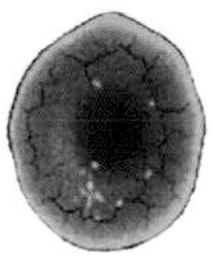	2 um 	500 nm 	200 nm 

Note: These sizes are approximate.
Sizes of actual objects will vary.

Red blood cell
Courtesy of CDC/
Janice Haney Carr

A strand of hair is around 0.4 mm (400 um) and is near the limit of what can be seen by the unaided human eye. Protists are around 200 um, plant cells are around 100 um, animals cells are around 20 um, and red blood cells are around 8 um. An *E. coli* bacterium is around 2 um (2000 nm), and most structures inside a cell are 200 nm and below.

3.3 The Light Microscope

The basic compound light microscope consists of an eyepiece or ocular lens, the magnifying (objective) lens, a stage for holding the sample, a light source, and various areas for adjusting focus, brightness, and magnification.

A light microscope works when light is passed through a sample and the lenses collect the light, bend the light to provide magnification, and separate the bent light to make details in the sample visible. In other words, a microscope magnifies a sample, resolves details in a sample, and creates contrast so a sample can be seen.

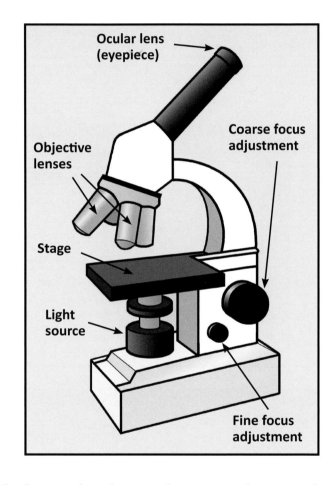

The magnification of a lens (how much a microscope can enlarge a sample) is expressed as numerical multipliers. A 2X magnification means that the lens doubles the size of the sample. In a 10X magnification the lens makes the sample 10 times larger, and a 100X magnification makes the sample 100 times larger. A modern compound microscope can magnify an object up to 1000 times!

Both the eyepiece and the objective lens magnify a sample. The eyepiece is the lens you look through, and the objective lens is the lens closest to the sample. The total magnification is the magnification of the eyepiece times the magnification of the objective lens. For example, if the eyepiece is 10X and the objective lens is 40X, the total magnification is 400X. The very smallest objects the unaided human eye can see are around 0.1 millimeter

(100 micrometers), so objects smaller than 100 micrometers need to be enlarged to be seen. This is what magnification does — it enlarges small objects so we can both see them and observe details.

However, magnification of a sample is not the only important feature of a microscope. More important than magnification is the resolution of the sample by the lens. Resolution is the smallest distance between two points on a sample that a lens can clearly define as being separate. Resolution is essentially the ability of a microscope to separate fine details on a sample.

For example, imagine two dots on a page. If we magnify the dots, we can make them appear bigger, but if we cannot separate them, they will appear to be a single, merged dot.

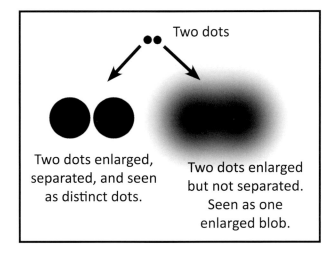

Two dots

Two dots enlarged, separated, and seen as distinct dots.

Two dots enlarged but not separated. Seen as one enlarged blob.

The resolution depends on the wavelength of light and the quality of the objective lens. The objective lens is the lens closest to the sample, and it is the only lens that is involved in providing resolution. The ocular lens (eyepiece) only magnifies what the objective lens resolves. In other words, if you use a higher magnification on the eyepiece but the objective lens cannot separate two points, the image will look big but will be blurry.

3.4 The Electron Microscope

Light microscopy is limited by the wavelength of visible light. Objects below about 300 nm cannot be resolved with a light microscope. This means that many small structures inside and on the surface of cells cannot be observed. Before the 1930s many scientists believed that it would never be possible to see things that are smaller than 300 nm. However, during the 1930s scientists began experimenting with electrons, and by the late 1950s the electron microscope had broken through the resolution limit of the light microscope.

The electron microscope uses a beam of electrons focused by a magnetic field much in the way a glass lens focuses a light beam. The "lens" for an electron microscope is called a solenoid, which is a coil of wire wrapped around the outside of a tube. When an electric current is passed through the wire, an electromagnetic field is created that can be used to control the electron beam.

The resolution of the electron microscope depends on how fast the electrons are traveling. The faster the electrons travel, the shorter their wavelength and the higher the resolution. With an electron microscope, samples can be magnified almost 2 million times!

Electron microscopes are very expensive pieces of equipment and are generally found only in specialized research labs. The electron microscope has a large vacuum chamber that houses an electron gun and magnetic lenses. The sample is inserted in the viewing chamber which is below the vacuum chamber, and a beam of electrons is directed from the electron gun toward the sample below. In a scanning electron microscope (SEM), an electron

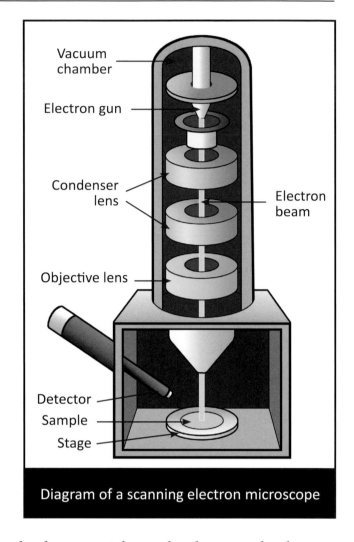

Diagram of a scanning electron microscope

beam is scanned across the sample surface, and a detector picks up the electrons that have been scattered during the scan. A computer then uses this information to create an image.

3.5 Scanning Probe Microscopes

How small are molecules? How small are atoms? How small is DNA, RNA, or a protein? The cells that make up living things are small, but even the smallest living cell is made up of billions of proteins, molecules, and atoms!

For a long time scientists could not observe small molecules found in cells or the atoms that make them. However, in the 1980s a new microscope technology was invented. This new device is called a scanning tunneling microscope, or STM. An STM is part of a family of microscopes called probe microscopes that make it possible to "see" small molecules and even atoms.

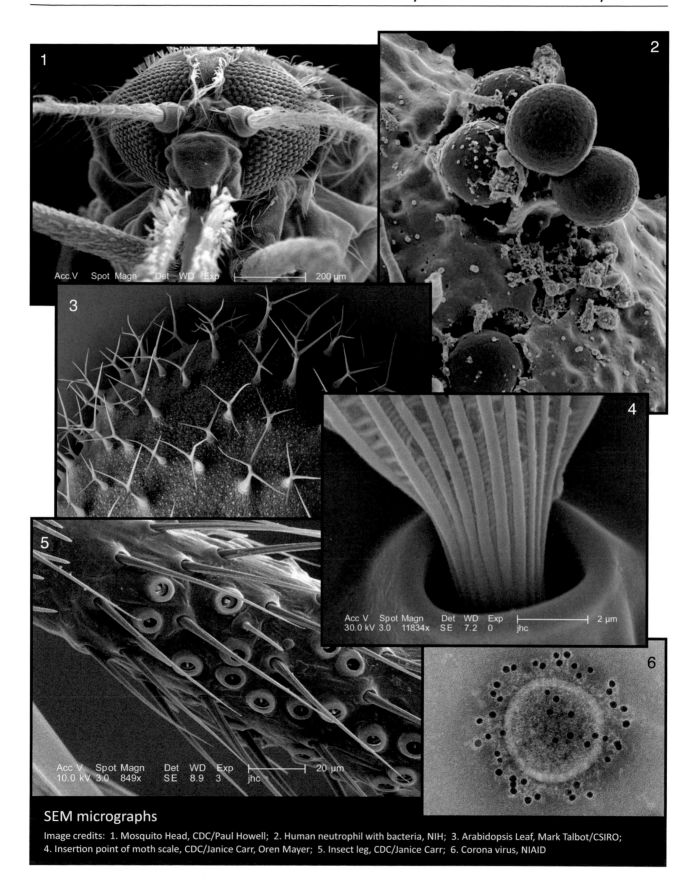

Acc.V Spot Magn Det WD Exp 200 µm

Acc V Spot Magn Det WD Exp
30.0 kV 3.0 11834x SE 7.2 0 jhc 2 µm

Acc V Spot Magn Det WD Exp
10.0 kV 3.0 849x SE 8.9 3 jhc 20 µm

SEM micrographs

Image credits: 1. Mosquito Head, CDC/Paul Howell; 2. Human neutrophil with bacteria, NIH; 3. Arabidopsis Leaf, Mark Talbot/CSIRO; 4. Insertion point of moth scale, CDC/Janice Carr, Oren Mayer; 5. Insect leg, CDC/Janice Carr; 6. Corona virus, NIAID

A scanning tunneling microscope is not a typical microscope. It does not work with light or lenses, and you don't look through it. In fact, when using an STM, you do not actually "see" the atoms, at least not in the way that you are looking at this page in front of you.

An STM works by "scanning" the surface of an object and then projecting an image of this surface onto a computer monitor or other screen. The STM has a metal probe called a stylus that actually does the scanning. The stylus is extremely sharp — it comes to a point that is only one atom wide!

The stylus is controlled by a computer and moves back and forth over the surface of the object that is being scanned. The stylus stays very close to the surface of the object with the gap between the tip of the stylus and the object being about as wide as one atom, or even smaller. The precision required to keep the stylus moving at the right distance from the scanning surface would not be possible without computers. As the stylus moves, it "picks up" electrons from the surface of the object. The electrons show where the atoms in the object are located. The STM electronically amplifies the signals created by these electrons, and a computer then interprets the signals, creating an image on a monitor.

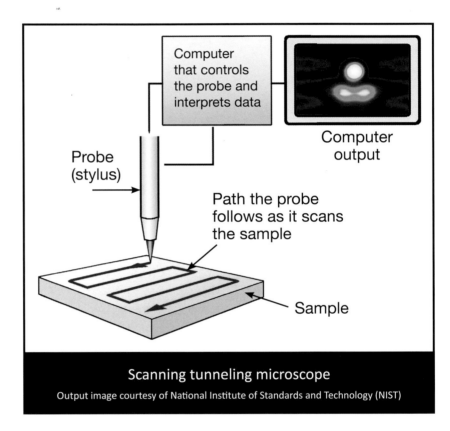

Computer that controls the probe and interprets data

Computer output

Probe (stylus)

Path the probe follows as it scans the sample

Sample

Scanning tunneling microscope
Output image courtesy of National Institute of Standards and Technology (NIST)

Xenon atoms written on a surface to spell IBM
Reprinted with permission from IBM Corporate Archives

An STM can produce phenomenal images of a surface, but it has another amazing function. An STM can be used to "grab" individual atoms! The computer controlling the STM can then move the atoms to specific locations. In 1990, researchers at IBM used an STM to grab individual xenon atoms. It took over 20 hours, but they were able to arrange 35 atoms into the letters I, B, and M to make the smallest company logo ever.

Since then, researchers have been discovering ways to move atoms around more quickly and how to make incredibly tiny structures, one atom at a time.

One of the drawbacks of the early scanning tunneling microscopes was that they could only

STM image of atoms: arsenic (yellow), manganese (red)

Courtesy of National Institute of Standards and Technology (NIST)

A structure made by an STM—Cobalt atoms on a copper surface

Image courtesy of National Institute of Standards and Technology (NIST)

be used to scan objects such as metals that conduct electricity easily. Therefore, they could not be used to create images of substances that were not conductors of electricity, such as plastics or living tissues. In the years since STMs were invented, several other types of probe microscopes have been developed. They work in slightly different ways, but all of these microscopes allow scientists to get an extremely close-up image of very small objects.

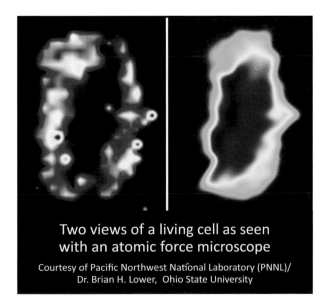

Two views of a living cell as seen with an atomic force microscope

Courtesy of Pacific Northwest National Laboratory (PNNL)/ Dr. Brian H. Lower, Ohio State University

One of these probe microscopes is the atomic force microscope (AFM) which can scan many different types of surfaces, including metals and nonmetals. Like an STM, an AFM stylus has a very sharp tip. But instead of picking up electrons to create an image like an STM does, an AFM can "see" atoms by just bumping into them (that is, by measuring the *force* between an atom and the tip of the probe).

Because everything is made of atoms, an AFM can "see" all kinds of materials, not just those that conduct electricity. The AFM has been used to image the surfaces of cells and observe small proteins in action.

3.6 Summary

- Different types of microscopes include the light microscope, the electron microscope, and a family of probe microscopes.

- A light microscope works when light passes through a sample and lenses collect the light, provide magnification of the sample by bending the light, and separate the bent light so that details in the sample are visible.

- The amount of magnification of a light microscope is the magnification of the ocular lens (eyepiece) times the magnification of the objective lens (the lens closest to the sample).

- The objective lens determines the resolution of a light microscope.

- Electron microscopes use a beam of electrons and a magnetic "lens" to image a sample.

- Probe microscopes, including the scanning tunneling microscope (STM) and atomic force microscope (AFM), scan the surface of a sample and can produce images of small molecules and atoms.

3.7 Some Things to Think About

- How do you think the invention of the microscope changed our understanding of cells and how they work?

- Describe the different metric units used in microscopy. Why are British units not used?

- Do you think a light microscope would work well for observing some types of samples but not others? Why or why not?

- Do you think there will be times when you want to view a sample at low magnification rather than high magnification or vice versa? Why or why not?

- Looking at the micrographs in this chapter, what features can you see that you would not be able to notice without using an scanning electron microscope?
  What do you observe that is surprising or unexpected?

- In what ways do you think scanning electron microscopes and scanning probe microscopes have profoundly changed our understanding of biology?

Chapter 4 The Chemistry of Life

4.1 Introduction

All things, including living things, are made of small units of matter called atoms and molecules. Atoms and molecules join to form cells, which are the basic building blocks of life. What makes a cell special is the high organization of atoms and molecules that work together through chemical reactions to keep the cell alive.

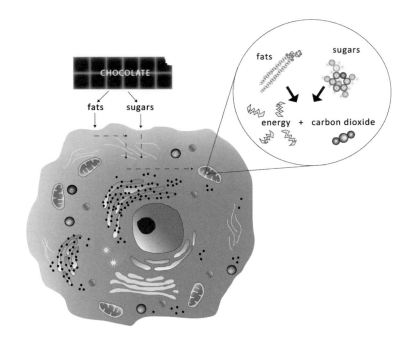

For example, in order for a cell to be alive, it must convert food into energy. To convert food into energy, food molecules are broken down into smaller pieces during a series of chemical reactions within the cell. When this happens, chemical bonds are created and destroyed and energy is released. These processes keep cells alive.

There are literally millions of chemical reactions going on in each cell at every moment. All of these chemical reactions are performed by millions of little molecules reacting with each other. When these chemical reactions stop, the cell dies.

4.2 Types of Atoms Inside Cells

There are over 100 different atoms, but the majority of biological molecules are made up of just six different atoms. This set of atoms is called the HCNOPS group (pronounced: aich-see-nops) and includes hydrogen, carbon, nitrogen, oxygen, phosphorus, and sulfur.

One of the special features that distinguishes living things from nonliving things is that living things have biological molecules. Most biological molecules are made of the five atoms in the HCNOPS group.

4.3 Types of Biological Molecules

There are many different types of biological molecules, including large biological molecules made up of millions of atoms and small biological molecules made up of only a few atoms. Some biological molecules are made up of smaller molecules that are hooked together to form long chains, and these molecular chains are grouped together in particular shapes.

Biological molecules perform different jobs inside cells. Some biological molecules provide energy for chemical reactions. These are called energy molecules. Other biological molecules are used to hold different parts of the cell together. These are called structural molecules. Some molecules move other molecules, break down unwanted molecules, and make molecules. These molecules are all molecular machines. And some molecules give the cell instructions for how to grow and when to die. These molecules are called information molecules.

All of these different types of biological molecules work together to keep cells alive, make them grow, process energy, and eventually tell the cell when to die.

4.4 Energy Molecules

Energy molecules play an important role in many different biological processes. One of the most important jobs that molecules perform inside cells is storing and transferring energy. In order for a cell to use glucose, the cell must have a way to store and transfer the energy it gets from the glucose molecules. To store and transfer energy, cells use special energy molecules.

The most important energy molecule inside cells is called adenosine triphosphate, or ATP. The ATP molecule gets its name because it has an adenosine group attached to three phosphate groups. A phosphate group is a cluster of phosphorus, oxygen, and hydrogen molecules. In an ATP molecule, the energy is stored in the phosphate bonds. A phosphate bond is the bond between two phosphate groups.

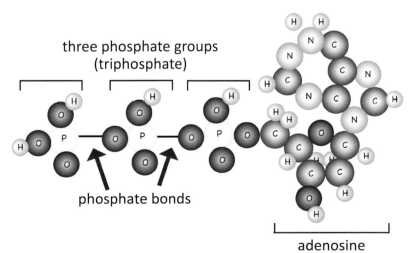

three phosphate groups
(triphosphate)

phosphate bonds

adenosine

When the phosphate bonds are broken during a chemical reaction, energy is released. Energy can be put back into the molecule when the phosphate bonds form again. The phosphate bonds can be broken and created over and over again, so the cell can store and release energy as often as needed. It is useful to think of ATP as being like a little rechargeable battery that stores and releases energy as the cell needs it.

4.5 Structural Molecules

Cells also have molecules that hold the cell together. These molecules are called structural molecules.

Plant cells, for example, have a stiff outer wall that helps the plant stand upright. This stiff outer wall is made of cellulose. Cellulose is a structural molecule composed of millions of glucose molecules.

Cellulose

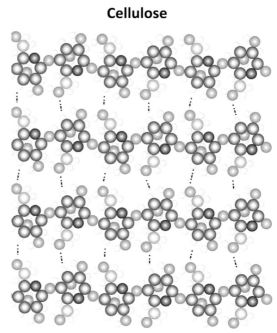

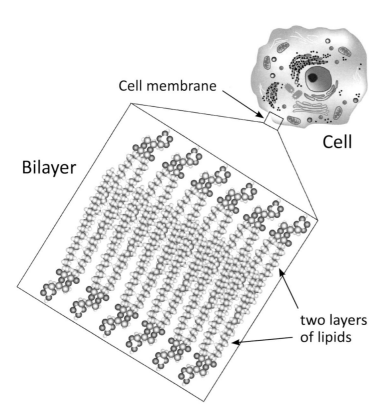

Cell membrane

Bilayer

Cell

two layers of lipids

An animal cell is surrounded by a cell membrane. The cell membrane is made of a type of fat molecule called a lipid. The lipid molecules in a cell membrane form two layers that together are called a lipid bilayer. The lipid bilayer is a structural molecule made of lipids, and it holds a cell together.

Structure is provided to cells by the molecules in both cellulose and lipid bilayers. Both types of molecules are structural molecules.

4.6 Molecular Machines

Cells perform a variety of different jobs, and to do those jobs cells use molecular machines. Molecular machines are specialized molecules that can cut other molecules apart and glue them back together. Molecular machines can also move other molecules around inside the cell or transport them outside the cell. Some molecular machines read molecules, others transcribe molecules from one molecular language to another, and others move molecules from one end of the cell to the other end.

Kinesin is a molecular machine that moves cargo around on a molecular "road" within the cell.

Kinesin "walks" along a molecular "road" and carries cargo

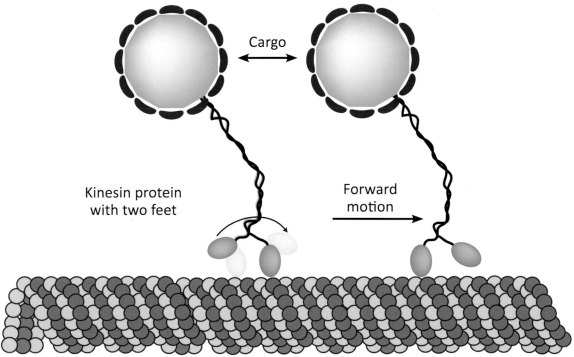

Molecular "road" (microtubules)

Molecular machines are made mostly of proteins. Proteins are a special type of biological molecule made of long chains of molecules (polymers) that can fold into a variety of structures. The structure of the folded protein is important. If a protein does not fold properly, it won't function. Proteins can perform a wide array of important jobs for the cell because they can fold into so many different shapes.

4.7 Information Storage and Transfer

Deoxyribonucleic acid (DNA) and ribonucleic acid (RNA) are special types of biological molecules that are used by the cell to store and transfer information. DNA and RNA store and transfer all the information a cell needs to grow, divide, make proteins, and eventually die. The information in DNA and RNA molecules is like the code in a computer. This code tells the cell when to grow, how to convert food into energy, when to stop growing, and when to die. The cell uses molecular machines to read the DNA code and make sense of it so the cell can know what functions to perform.

Deoxyribonucleic acid (DNA) is a polymer that is made of nucleotides which are made of two parts: nucleic acid bases and ribose sugars. The bases are connected to the sugars, and the sugars are connected to each other. RNA is similar to DNA but lacks an oxygen group on the sugar. RNA also folds differently than DNA.

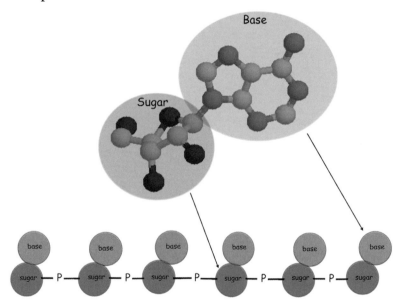

4.8 Chemical Reactions in Cells: Metabolism

Metabolism is the term used to refer to all of the chemical processes your body uses to stay alive. The word metabolism comes from the Greek word *metabole* which means "to change or overthrow." In biology, the word metabolism describes all of the chemical reactions living things use to change food into energy and other materials in order to live and grow.

There are literally millions of chemical reactions happening in your body every minute. Your cells are constantly making and destroying molecules, which converts them from one form to another, and cells are constantly creating and using energy.

Today we know a great deal about the chemical reactions cells use for metabolism. Chemical reactions in a cell follow a certain order, with one chemical reaction leading to another. This order is called a metabolic pathway. There is a different metabolic pathway for each function the cell is performing.

Two of the most important metabolic processes cells perform are the conversion of food into energy and the conversion of energy into food. The conversion of food into energy is called glycolysis. The conversion of energy into food is called photosynthesis.

Animals use glycolysis to convert the food they eat into energy, and plants use photosynthesis to convert energy from the Sun into food. Both glycolysis and

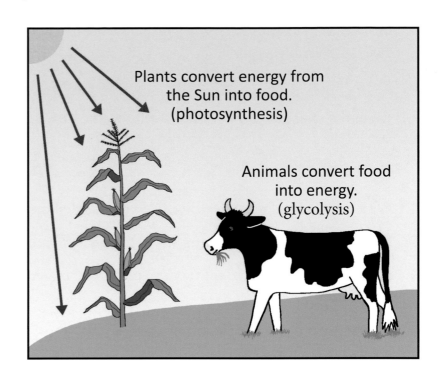

Plants convert energy from the Sun into food. (photosynthesis)

Animals convert food into energy. (glycolysis)

photosynthesis are metabolic pathways with many chemical reactions linked together. And photosynthesis and glycolysis depend on each other. Without photosynthesis, glycolysis cannot happen and without glycolysis, photosynthesis cannot happen. Because these processes depend on each other, the organisms that perform them depend on each other. Plants depend on animals and animals depend on plants!

4.9 Summary

- Living things are made of atoms and molecules that are highly organized and perform millions of chemical reactions every moment.

- Most biological molecules are made of six different atoms called the HCNOPS group which includes hydrogen, carbon, nitrogen, oxygen, phosphorus, and sulfur.

- There are different types of biological molecules that perform different functions inside cells.

- Energy molecules, such as ATP, are like little batteries that store and release energy.

● Structural molecules, such as cellulose and lipid bilayers, give cells the structure they need to function properly.

● Molecular machines are mostly proteins that can read, cut, transcribe, and move molecules inside a cell.

● Molecules such as DNA share and transfer information.

● All of the chemical processes in a cell together are called metabolism.

4.10 Some Things to Think About

● Why do you think chemical reactions are needed for cells to stay alive?

● Why do you think living things need to have different molecules than nonliving things?

● How do you think different types of biological molecules are able to perform different functions?

● Why do you think it is important to a cell that phosphate bonds can be broken and created over and over again?

● Why do you think there is a need for different types of cell membranes?

● How important do you think proteins are for your body? Why?

● Do you think that since DNA and RNA have different structures, they are able to perform different tasks? Why?

● Do you think it is important that chemical reactions in cells happen in a certain order? What do you think might happen if the reactions took place in a different order? Why?

Chapter 5   Cells—The Building Blocks of Life

5.1 Introduction

Recall that all living things are made of a complex and highly ordered arrangement of atoms and molecules that fit together to form cells. Cells are the building blocks of all living things.

Each cell is like a little factory. All of the molecules in the cell have special jobs to do. There are many different kinds of molecules inside cells. Most of the big molecules are proteins, but there are also sugars and nucleic acids. Small molecules such as water and salt are also in cells.

The cell "knows" where all of the molecules are and how many molecules are doing work. This way the cell always has just the right number of molecules working in a particular area, and there are not too many or too few molecules.

The molecules in a cell never rest. They are always working. If they stop working, the cell can no longer live. When cells get old or are damaged, the molecules cannot work, and the cell dies. When a cell dies, all the parts of the cell break down into smaller molecules. The molecules from dead cells are used again to make new molecules for different cells. Eventually, all cells in all living things die.

5.2 Types of Cells

There are three major types of cells that correspond to the three main taxonomic domain divisions. Organisms in the domain Bacteria have bacterial prokaryotic cells, organisms in the domain Archaea have archaeal prokaryotic cells, and organisms in the domain Eukaryra have eukaryotic cells. All three cell types have some similarities and some differences.

All three cell types are similar because they all are made of atoms and molecules. Also, all three types of cells act as little chemical factories, making the molecules and structures needed by the cells and breaking down the molecules that the cells don't need.

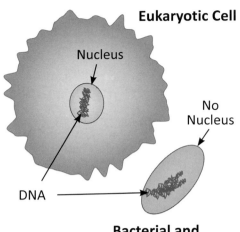

Eukaryotic Cell

Nucleus

No Nucleus

DNA

Bacterial and Archaeal Cells

However, these three cells types are different from each other in significant ways. For example, eukaryotic cells have a nucleus. A nucleus is a small "sack" that is inside the cell and holds the DNA. Neither bacterial nor archaeal cells have a nucleus to hold their DNA.

Both eukaryotic cells and bacterial cells have plasma membranes that are different from the plasma membrane of archaeal cells. A plasma membrane is a thin, soft, greasy film that surrounds the cell and controls what goes into and out of the cell. Bacterial and eukaryotic cells have a glycerol ester plasma membrane, and the archaea have a slightly different plasma membrane called a glycerol ether plasma membrane. Glycerol ester and glycerol ether plasma membranes get their names from the kind of molecules they are made of.

Bacterial and archaeal cells do not have organelles, but eukaryotic cells do. (An organelle is like a little organ because it is a structure that performs specific functions inside a cell.) Both archaea and eukarya have eukaryotic ribosomal proteins, and bacteria have prokaryotic ribosomal proteins. (A ribosomal protein is the molecular machine that makes proteins.) Although there are similarities between these three cell types, there are enough differences that three different domains are needed.

Comparison of Cell Types		
Bacteria Prokaryotic Cells	**Archaea** Archaeal Cells	**Eukarya** Eukaryotic Cells
Nucleus No	No	Yes
Organelles No	No	Yes
Plasma Membrane Glycerol ester	Glycerol ether	Glycerol ester
Ribosomal Proteins Prokaryotic	Eukaryotic	Both eukaryotic and prokaryotic

5.3 Bacterial Prokaryotic Cells

Bacteria are single-celled organisms. The word bacterium comes the Greek word *bakterion* which means "small staff or rod" and was used because the first bacteria observed were rod-shaped.

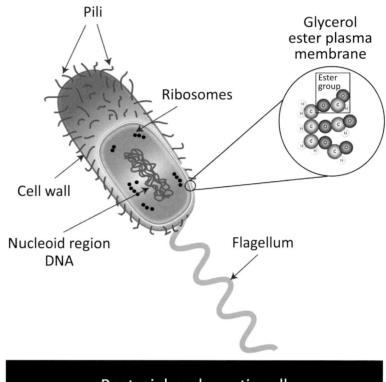

Bacterial prokaryotic cell

The word prokaryote comes from the Greek word *pro*, which means "before" and *karyon* which means "kernel." So prokaryote means "before kernel." This term refers to the fact that prokaryotes do not have a nucleus, the small "sack" inside the cell that holds the DNA in eukaryotes. Instead, the DNA is kept in a region called the nucleoid. A nucleoid is a central region in the cell that is not physically separated from the rest of the cell by a membrane.

Bacterial prokaryotic cells are surrounded by a cell wall and a glycerol ester plasma membrane. The cell wall is rigid, like a coat of armor, and protects the cell from being broken.

Many prokaryotic cells have flagella. These are long or short "whips" that help the cell move around. A flagellum is connected to a complicated molecular motor that twirls around with great speed and propels the cell in all directions. Some prokaryotes also have pili. Pili are long "threads" that help the cell stick to surfaces and to other cells.

5.4 Archaeal Prokaryotic Cells

Archaea are single-celled organisms. The word archaea comes from the Greek word *archein* which means "the first" or "to rule." Some scientists believe that archaeal cells were the first type of cells to exist. However, since archaeal cells have some similarities with eukaryotic cells, it isn't perfectly understood what the first cells may or may not have been.

Like bacterial prokaryotic cells, archaeal cells don't have a true nucleus or organelles. Archaeal cells have a glycerol ether plasma membrane, and some have flagella and pili. Most have a cell wall.

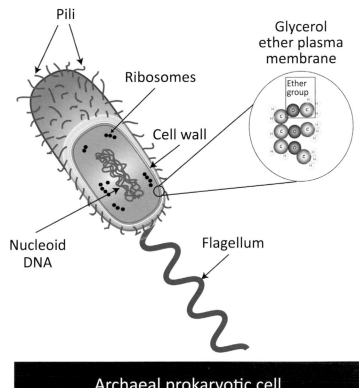

Archaeal prokaryotic cell

5.5 Eukaryotic Cells

The word eukaryote comes from the Greek word *eu* meaning "true" and *karyon*, meaning "kernel." Eukaryotic cells have a nucleus in the central part of the cell. Eukaryotic cells are usually much bigger and more complicated than either bacterial prokaryotic cells or archaeal prokaryotic cells. All living things in the domain Eukarya and the kingdoms Plantae, Animalia, Fungi, and Protista have eukaryotic cells.

There are two main types of eukaryotic cells—plant cells and animal cells. Because plant cells are eukaryotic cells, they have a nucleus, a glycerol ester plasma membrane, and organelles. Organelles function like little organs inside a cell. They are not true organs because true organs, such as the heart, are made up of many cells and an organelle is contained within an individual cell. But because each type of organelle performs specific functions, organelles can be thought of as little organs inside cells.

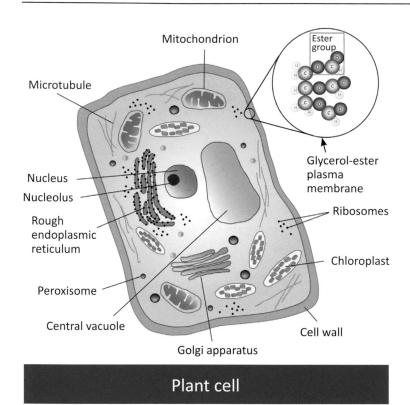

Mitochondrion

Microtubule

Nucleus
Nucleolus
Rough endoplasmic reticulum

Peroxisome

Central vacuole

Golgi apparatus

Ester group

Glycerol-ester plasma membrane

Ribosomes

Chloroplast

Cell wall

Plant cell

Many plants make their food from the Sun and have a type of organelle called a chloroplast. By using chloroplasts, many plants can convert the Sun's energy into food through photosynthesis. Plant cells have other organelles, including mitochondria (small factories that make ATP energy molecules), microtubules to move things from place to place, and a Golgi apparatus where proteins are modified, shipped, and stored.

Animal cells have many features similar to plant cells, and they also have important differences. Animal cells do *not* have a cell wall. Instead they are surrounded only by the plasma membrane. Also, animal cells don't have chloroplasts and cannot make their own food with sunlight.

Animal cells do have many other organelles that are the same as those found in plant cells. Animal cells have a nucleus, mitochondria, a rough endoplasmic reticulum, and microtubules that move things from place to place in the cell. Both also have a Golgi apparatus that behaves like a shipping and receiving area for the cell.

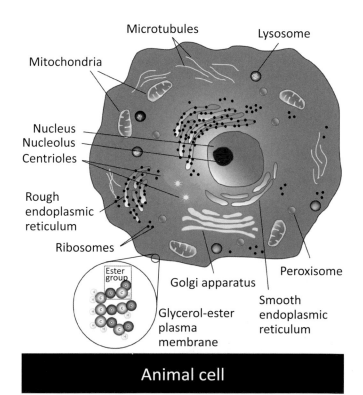

Microtubules

Lysosome

Mitochondria

Nucleus
Nucleolus
Centrioles

Rough endoplasmic reticulum

Ribosomes

Ester group

Golgi apparatus

Glycerol-ester plasma membrane

Peroxisome

Smooth endoplasmic reticulum

Animal cell

A Microscopic View of Some Eukaryotic Cell Structures

Scientists stain cells in different ways to show different structures.

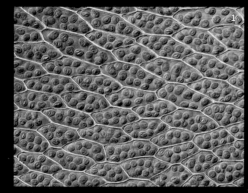

Chloroplasts in cells of a moss plant

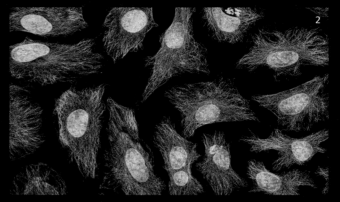

Human cells - DNA is blue, microtubules are magenta

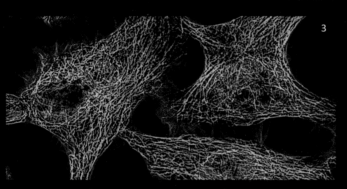

Human cells - DNA is blue, microtubules are cyan, nuclei are dark blue

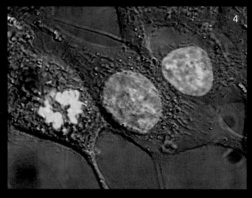

Human cells - DNA is blue. Cell on left is beginning to divide

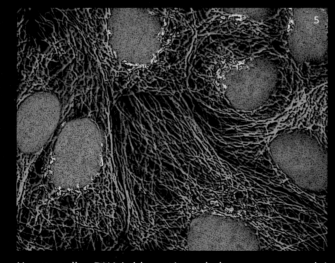

Human cells - DNA is blue, microtubules are green, nuclei are dark blue, Golgi apparatus are orange

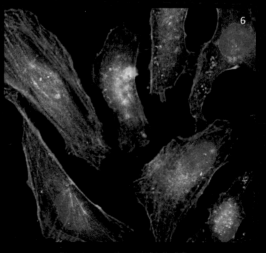

Human cells - DNA is blue

Photo credits: 1. Des Callaghan, CC BY SA 4.0, 2-5. NIH, Public Domain; 6. Gerry Shaw, CC BY SA 3.0

5.6 Cell Division

How does your body grow? As you grow, do your cells just get bigger or do they make more cells? Can you think of some advantages of being able to make more cells instead of just making bigger cells?

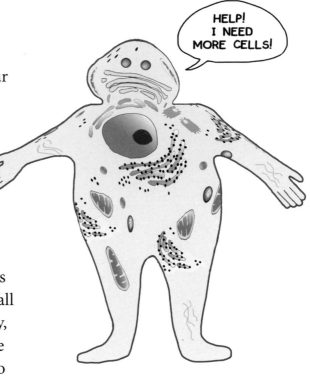

For you to grow and live a long life, your cells need to divide to make new cells. There are different reasons why our cells need to make new cells. The body needs to make different kinds of cells in order to have all the different organs it needs to function properly, and new cells are needed to grow and form these organs. New cells are needed to repair damage to the body. Your body makes new skin cells to heal cuts and scrapes. If your body did not make new skin cells, none of these injuries would heal. When you break a bone, your body can generate new bone cells to mend the break. Also, the lining of your small intestine has new cells replacing the old cells very rapidly, which keeps your digestive system working well.

The cells in your body are constantly replacing themselves. Old cells die and new ones are formed. When cells divide and make new cells, the DNA is copied so the new cells will have the correct instructions to function properly.

5.7 Bacterial and Archaeal Cell Division

Both bacterial prokaryotic cells and archaeal prokaryotic cells divide by binary fission. Binary means two parts, and fission means splitting. Binary fission literally means splitting into two parts.

When it is time for a prokaryotic cell to divide, the cell begins to elongate and to duplicate its DNA. As the DNA is duplicated and separated, the cell is also growing bigger. When it is almost twice its normal size, the middle of the cell begins to pinch together, dividing the cell

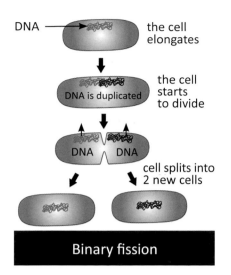

Binary fission

in two. The cell wall and plasma membrane grow down the middle of the cell, and when a new wall has formed, the cell divides into two new cells.

Under good growth conditions, bacteria will often grow, duplicate DNA, and divide almost constantly. If you start with one bacterium in ideal conditions, after a few hours you can have many millions of bacteria!

5.8 The Eukaryotic Cell Cycle: Mitosis

A eukaryotic cell goes through a much more complicated process with a series of stages that together make up its cell cycle. When eukaryotic cells divide to repair or grow, they divide by a process called mitosis.

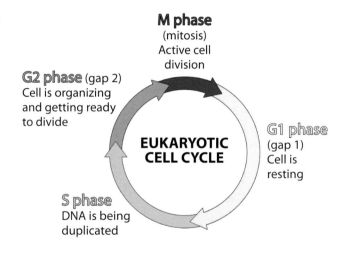

The cell cycle of mitosis has four main steps. The first step is called the G1 phase. G stands for gap, and this is the first gap, or space, between cell divisions. During this phase the cell has finished dividing, and the cell is resting, possibly getting bigger but not actively doing anything to prepare for the next division. Cells can stay in this phase for a short time or for a very long time.

S phase: S stands for synthesis, which means "making." In S phase the cell is beginning to prepare to divide by copying all of its DNA. At the end of S phase, there will be two copies of all the DNA.

G2 phase: During the second gap phase, the cell is finishing preparations to divide. It organizes some of the machinery that it will need to divide.

During mitosis (M phase) the two copies of DNA are pulled to opposite sides of the cell and the cell divides. After mitosis the cell goes back to the G1 phase and the process starts again.

5.9 Summary

● All living things are made of cells.

● There are three main types of cells: bacterial prokaryotic cells, archaeal prokaryotic cells, and eukaryotic cells.

● Prokaryotic cells do not have a nucleus or organelles and eukaryotic cells do.

● To live and grow, cells need to divide to make new cells.

● Bacterial prokaryotic cells and archaeal prokaryotic cells divide by binary fission.

● When they need to divide to repair or grow, eukaryotic cells divide by a process called mitosis.

5.10 Some Things to Think About

● What do you think would happen if the atoms and molecules in a cell were not in a highly ordered arrangement?
What if they could do whatever they wanted to?

● Do you think all cells perform most of the same functions as each other in order to live? Why or why not?

● Do you think it is possible to look at one cell and tell whether it comes from a plant or a bacterium? Why or why not?

● Why do you think many prokaryotic cells have a flagellum?

● List some features that are the same in both plant and animal cells and some features that are different.
Why do you think plant and animal cells are different?

● What do you think your body would be like if it were made of only one cell?

● Explain binary fission in your own words.

● Explain in your own words what happens during mitosis.

Chapter 6 Viruses, Bacteria, and Archaea

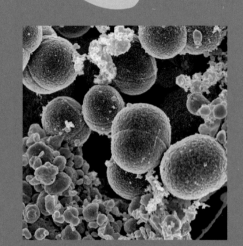

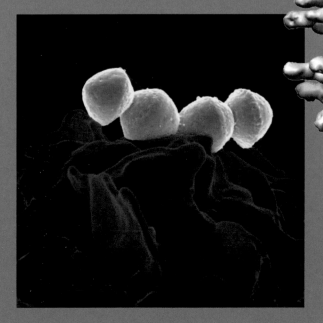

6.1 Introduction

As we saw in the last several chapters, all living things are made of atoms and molecules, and all living things assemble these atoms and molecules into small units called cells. Cells are the basic units of life. In the last chapter we took a close look at cells, the different types of cells, and how cells divide. In this chapter we will take a look at some of the smallest living things: viruses, bacteria, and archaea.

6.2 Viruses

If you start sneezing and two days later you feel achy and tired with a sore throat and a stuffy nose, you've probably been hit with a virus.

Viruses are odd, and many scientists still don't know how to classify them. Are viruses really alive or are they just some proteins and DNA codes that infect living things? Because it isn't clear how to catalog viruses, they are not in a domain and don't have their own kingdom.

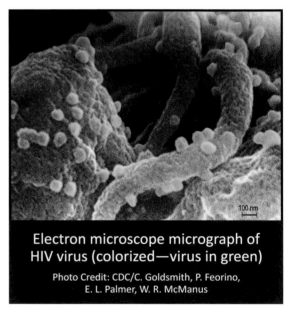

Electron microscope micrograph of HIV virus (colorized—virus in green)

Photo Credit: CDC/C. Goldsmith, P. Feorino, E. L. Palmer, W. R. McManus

Viruses are often classified according to the type of organism they infect. For example, Bunyaviruses are grouped together because they infect plants and animals, and Totiviruses are in a different group because they infect protists and fungi.

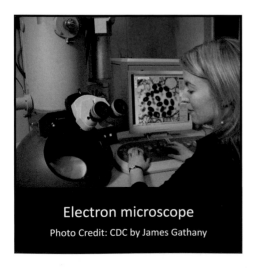

Electron microscope

Photo Credit: CDC by James Gathany

Viruses don't have all the features that bacterial, archaeal, or eukaryotic cells have. Viruses don't have a nucleus, a cell wall, or a cell membrane. Viruses are technically not cells but small sacks containing DNA or RNA that are surrounded by a tough outer coat made of protein. Since most viruses are much smaller than either bacteria or archaea and are too small to be visualized with a regular light microscope, an electron microscope is needed to see them. Recall that an electron microscope uses electrons instead of light to visualize very small particles.

Viruses can be spiral in shape and look like a tightly coiled garden hose, or they can be icosahedral and shaped like a soccer ball.

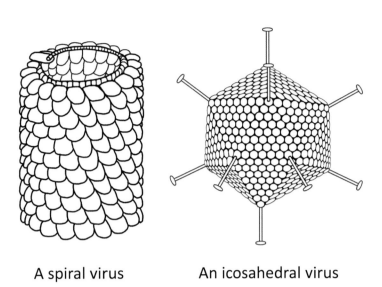

A spiral virus   An icosahedral virus

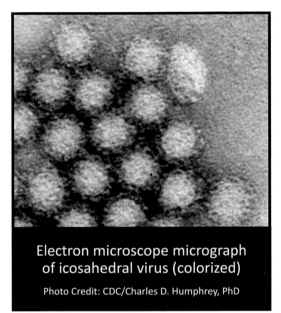

Electron microscope micrograph of icosahedral virus (colorized)

Photo Credit: CDC/Charles D. Humphrey, PhD

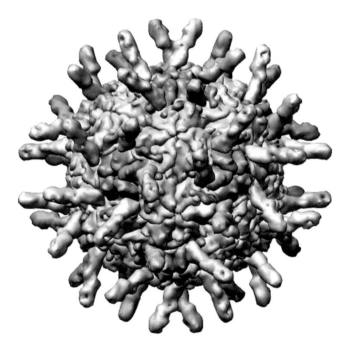

Illustration of Human Rhinovirus 16

Reference Protein Data Bank structural studies of two rhinovirus serotypes complexed with fragments of their cellular receptor, Kolatkar, P. R., Bella, J., Olson, N. H., Bator, C. M., Baker, T. S., Rossmann, M. G., Journal: (1999) EMBO J. 18: 6249-6259]

The group of viruses that causes most common colds is the rhinovirus group. There are actually three different categories of rhinoviruses: Rhinovirus-A, Rhinovirus-B, and Rhinovirus-C. Rhinoviruses-A and B cause most of the less severe common cold symptoms in humans, and Rhinovirus-C is believed to cause more severe cold symptoms.

When a rhinovirus is interacting with human proteins and cells, it looks like a soccer ball with spikes. Most rhinoviruses cause mild to severe infections in the lungs with occasional headaches and sometimes fever.

6.3 Bacteria

Bacteria have gotten a bad name because many bacteria make humans, plants, and animals sick. But not all bacteria are harmful. It may (or may not) make you feel better to know that there are lots of bacteria that don't hurt us, and some bacteria even help us. Scientists guess that there may be something like 100,000,000,000,000,000 (100 quadrillion) bacteria that live on and in each one of us!

In fact, bacteria live everywhere. People have found bacteria almost everywhere they have looked. As scientists have learned more about these tiny organisms, we have been able to use this knowledge to cure disease, grow food, and clean up our world.

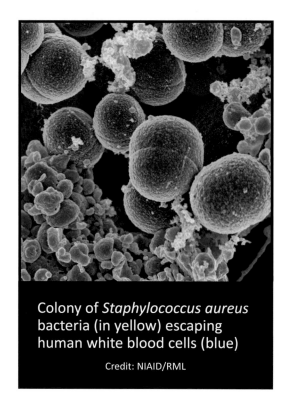

Colony of *Staphylococcus aureus* bacteria (in yellow) escaping human white blood cells (blue)

Credit: NIAID/RML

Bacteria are single-celled organisms with a prokaryotic cell type. Recall that bacteria have no nucleus and their DNA is "loose" in the cell, located in a centralized place called the nucleoid. Single bacteria are too small to see without a microscope, but a huge group of bacteria, called a colony, can be easily seen with the unaided eye.

Bacteria reproduce fast. Some take only ten minutes to grow and divide. If you started with one, after ten minutes you would have two; after twenty minutes, four; after thirty minutes, 8; after forty minutes, 16; after fifty minutes, 32; and after an hour, 64. If they have the right growing conditions, after a few hours or days there are so many bacteria that you can see the colony without using a microscope or magnifying glass. It's a good thing bacteria can't reproduce that quickly forever. They need the right conditions to grow so fast. If a colony gets too big, it can run out of food. Also, bacteria need moisture and don't thrive on dry surfaces.

When people boil water to make it safe to use, they destroy the bacteria cells. Bacteria can be killed by very high temperatures, like the temperature of boiling water (100° C, 212° F). Very low temperatures slow the rate at which bacteria divide. This is why refrigeration allows food to last for a longer period of time. However, low temperatures do not generally kill bacteria.

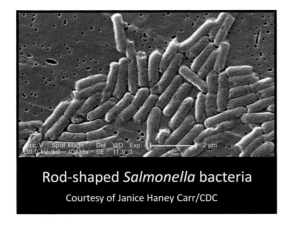

Rod-shaped *Salmonella* bacteria

Courtesy of Janice Haney Carr/CDC

6.4 Shapes of Bacteria

Bacteria come in many different sizes and shapes. The three most common shapes are rods, spheres, and spirals, although some bacteria can be star-shaped or even rectangular.

E. coli is an example of a rod-shaped bacterium. Rod-shaped bacteria are called bacilli. Most bacilli appear as single rods. Some bacilli form pairs after they divide and are called diplobacilli. Others form a chain and are called streptobacilli.

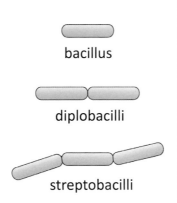

bacillus

diplobacilli

streptobacilli

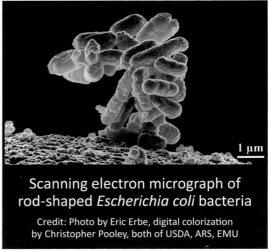

Scanning electron micrograph of rod-shaped *Escherichia coli* bacteria

Credit: Photo by Eric Erbe, digital colorization by Christopher Pooley, both of USDA, ARS, EMU

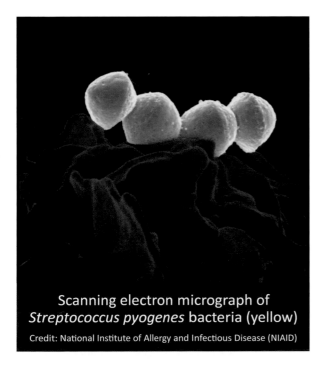

Scanning electron micrograph of *Streptococcus pyogenes* bacteria (yellow)

Credit: National Institute of Allergy and Infectious Disease (NIAID)

Sphere-shaped bacteria are called cocci, from the Greek word *kokkus* which means "grain, seed, or berry." Most cocci are round, but they can also be elongated or flattened on one side. After cocci divide, they can remain in pairs, form long chains, or form clusters. When you have strep throat, it is an infection of *Streptococcus* bacteria.

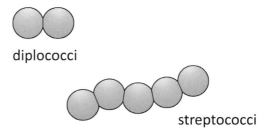

diplococci

streptococci

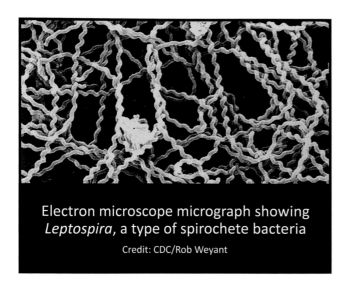

Electron microscope micrograph showing
Leptospira, a type of spirochete bacteria

Credit: CDC/Rob Weyant

There are three different types of spiral bacteria called vibrio, spirillum, and spirochete. Vibrio bacteria are almost like rods, but are curved. Spririlla are fairly stiff and look like a short corkscrew. Spirochetes are long and thin with flexible bodies, and they look like a long corkscrew.

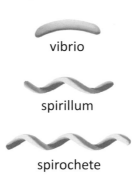

vibrio

spirillum

spirochete

6.5 Archaea

Arahaea are similar to bacteria in size and shape, but they are not identical and therefore have their own domain. Scientists first found archaea in environments where nothing else could survive. Three types of extreme environments where archaea can live are places without oxygen, places with extreme amounts of salt, and places that are very hot.

San Francisco Bay salt ponds
with colorful halophiles

Credit: Grombo Permission: GFDL, cc-by-sa-2.5. 2.0, 1.0

Methanogens are a type of archaea that cannot survive where there is oxygen. These archaea use carbon dioxide, nitrogen, and hydrogen for energy and give off methane. Methane is a gas that stinks! Many stinky places smell of methane — swamps, marshes, and cows' guts, for example. Cows need archaea to help them digest their food, and methane is made during the digestive process. When plants in a swamp die and the decomposing bacteria use up all the oxygen, it is the perfect environment for the methanogens. For this reason, people can use methanogens to treat sewage (waste water). Because methane can be burned, scientists are experimenting with ways to collect methane and use it for fuel.

Another group of archaea are the halophiles. Their name comes from the Greek words *hals* which means "salt" and *philein* which means "to love," so the halophiles love a salty environment. Halophiles live in bodies of water that have lots of extra salt; for example, the Dead Sea.

Thermoacidophiles like it hot! Thermoacidophiles are "hot acid lovers." These archaea live in the hot water right over volcanic areas, like deep sea vents where cracks in the Earth's crust spew out hot seawater full of minerals. The thermoacidophiles use these minerals for food.

Deep sea vents at the Champagne Vent Site near Japan

Credit: NOAA
(oceanexplorer.noaa.gov/gallery)

6.6 Summary

- Viruses, bacteria, and archaea are some of the smallest living things.

- Viruses can be spiral or icosahedral in shape. They are difficult to classify, and some scientists don't think viruses should be considered to be "alive."

- Bacteria live everywhere and can reproduce rapidly.

- Bacteria have three main shapes: rod, sphere, and spiral.

- Archaea were once grouped with bacteria but are different enough to have their own domain.

- Archaea can live in extreme environments.

6.7 Some Things to Think About

- How many cells is the smallest living thing made of?

- Why can't scientists classify viruses?

- What are some ways that you could slow down or stop bacteria from multiplying? How could this knowledge be helpful?

- Which type of bacteria do you find most interesting? Why?

    bacillus • diplobacilli • streptobacilli • diplococci • streptococci • vibrio • spirillum • spirochete

- Can you think of some other types of extreme environments where archaea might be able to live?

Chapter 7 Protists

2 microns

Photo credits: See end of chapter

7.1 Introduction

Protists, sometimes called protozoa, are organisms that are like both plants and animals. Protists are in the domain Eukarya and have their own kingdom called Protista. The word Protista comes from the Greek *protos* which means "first." Although it is not likely that protists were among the very first life forms to appear on the planet, they are some of the oldest organisms that have been found in the fossil records.

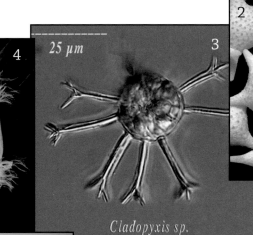

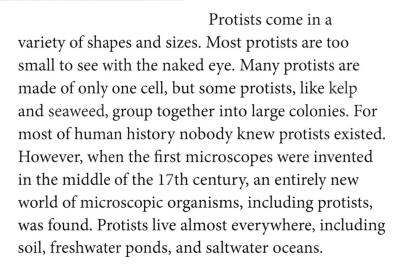

Cladopyxis sp.

Protists come in a variety of shapes and sizes. Most protists are too small to see with the naked eye. Many protists are made of only one cell, but some protists, like kelp and seaweed, group together into large colonies. For most of human history nobody knew protists existed. However, when the first microscopes were invented in the middle of the 17th century, an entirely new world of microscopic organisms, including protists, was found. Protists live almost everywhere, including soil, freshwater ponds, and saltwater oceans.

1. Dinoflagellate, Courtesy of CSIRO; 2. Formanifora, Courtesy of Psammophile (CC BY SA 3.0); 3. Dinoflagellate, Courtesy of Dr. John R. Dolan, Laboratoire d'Oceanographique de Villefranche; Observatoire Oceanologique de Villefrance-sur-Mer; 4. *Didinium nasutum*, Courtesy of Gregory Anitpa, San Francisco State University; 5. Giant kelp, Courtesy of Claire Fackler, CINMS/ NOAA

7.2 Classification

It is unknown how many species of protists exist, but estimates range from 36,000 to 200,000, many of which have not yet been discovered. Although protists are classified in the single kingdom, Protista, they vary in structure and function more than any other group of organisms. Because this group is so diverse, there are several different classification systems for protists. In this text we will focus on four main groups depending mostly on how they move. These group are the ciliates, the flagellates, the amoebae, and the spore-forming protists.

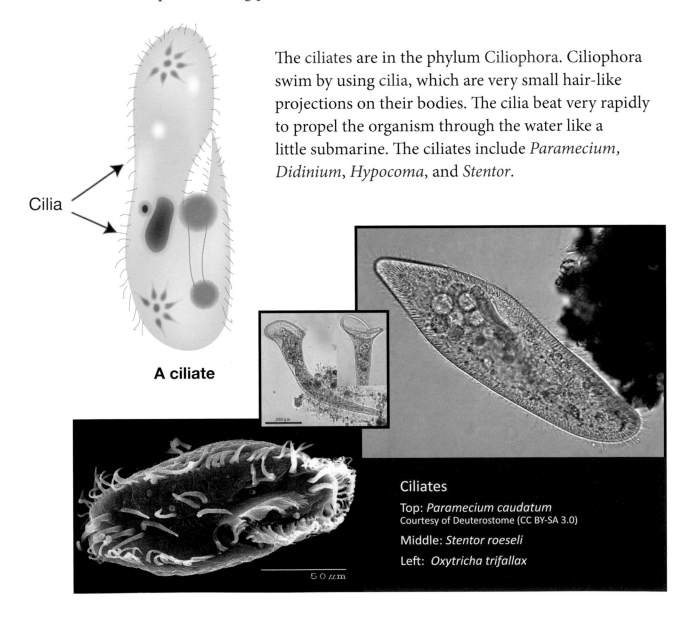

Cilia

The ciliates are in the phylum Ciliophora. Ciliophora swim by using cilia, which are very small hair-like projections on their bodies. The cilia beat very rapidly to propel the organism through the water like a little submarine. The ciliates include *Paramecium, Didinium, Hypocoma,* and *Stentor.*

A ciliate

Ciliates

Top: *Paramecium caudatum*
Courtesy of Deuterostome (CC BY-SA 3.0)

Middle: *Stentor roeseli*

Left: *Oxytricha trifallax*

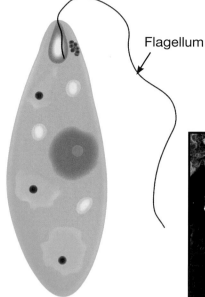

Flagellum

A Flagellate

There are several different phyla for the flagellates including Trichozoa, Euglenozoa, Dinozoa, Choanozoa, and Metamonada. Flagellates also swim, but instead of many short, hair-like projections, flagellates have one or more long whip-like flagella that extend from one end of their body. These whips propel the flagellates through the water much like the tail of a fish. Many flagellates have a thin outer covering called a pellicle. Flagellates can exist as single organisms or in colonies. Many flagellates are parasitic, living inside other organisms.

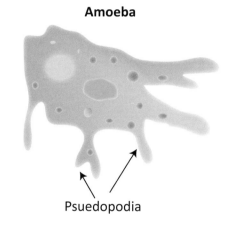

A dinoflagellate
Micrograph courtesy of CSIRO

Amoebae move very differently than the ciliates and flagellates. Amoebae do not swim or use flagella or cilia but instead crawl along surfaces by extending and bulging the edges of their membranes. The portions of their membranes that stick out are called pseudopodia. *Pseudo* is Greek and means "false" and *podia* means "feet," so pseudopodia are "false feet." Once the pseudopodia are extended, the rest of the amoeba flows into them, pulling the amoeba forward. The process then begins again.

Amoeba

Psuedopodia

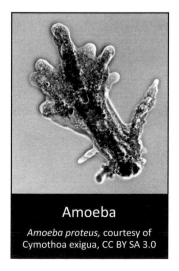

Amoeba
Amoeba proteus, courtesy of Cymothoa exigua, CC BY SA 3.0

In a microscope, the movement of an amoeba along the surface of a glass slide looks something like the following illustration.

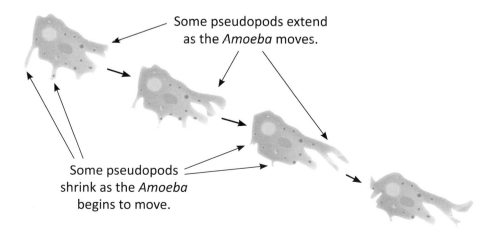

Some pseudopods extend as the *Amoeba* moves.

Some pseudopods shrink as the *Amoeba* begins to move.

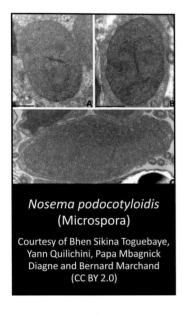

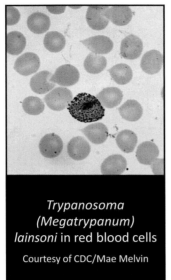

Nosema podocotyloidis
(Microspora)

Courtesy of Bhen Sikina Toguebaye,
Yann Quilichini, Papa Mbagnick
Diagne and Bernard Marchand
(CC BY 2.0)

*Trypanosoma
(Megatrypanum)
lainsoni* in red blood cells

Courtesy of CDC/Mae Melvin

Sporozoa, or spore-forming protists, include three major phyla — Apicomplexa, Microspora, and Myxosporidia (Myxospora). Sporozoans live as parasites within cells or organs of almost every kind of animal. Sporozoans do not have flagella, cilia, pseudopodia, or any other locomotion process. A sporozoan spends much of its life cycle unable to move by itself and passes from host to host in a protective capsule called a spore.

7.3 Photosynthetic Protists

Because most protists are single-celled organisms, they do not have the advantage of using tissues and organs to process food. Instead, they must gather food, digest nutrients, and eliminate wastes, all within a single cell. As a result, protists are much more complicated than cells of other eukaryotic organisms.

Some protists contain chloroplasts and use carbon dioxide, water, and the Sun's energy to make food by photosynthesis, just like plants do. Organisms that make their own food are called autotrophs. Autotroph comes from the Greek words *auto* which

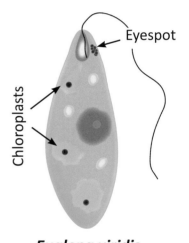

Euglena viridis

means "self" and *trophe* which means "food or nourishment," so an autotroph is an organism that is fed by itself.

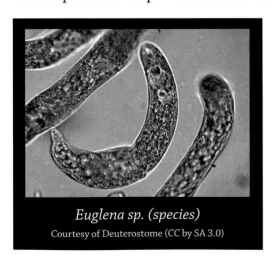

Euglena sp. (species)
Courtesy of Deuterostome (CC by SA 3.0)

Euglena viridis is an example of a photosynthetic protist. Euglena are found in freshwater streams and ponds, sometimes being so numerous that the water turns green. Because euglena depend on photosynthesis for food, it is important for them to be able to detect the sunny areas in a pond or stream. To detect light, Euglena have a small red spot toward the end of their body near the flagellum. This spot is called the eyespot or stigma. The stigma is a light sensitive area shaped like a shallow cup. This shape allows the euglena to detect sunlight only from a particular

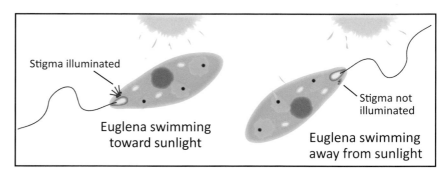

Stigma illuminated

Euglena swimming toward sunlight

Stigma not illuminated

Euglena swimming away from sunlight

direction. When the euglena is traveling toward the light, a small part in the base of the stigma is illuminated. When the euglena swims away from the light, the spot is no longer illuminated, and the euglena knows that it is no longer in the path of the sunlight. Using the stigma as a detector, the euglena can find the sunlight needed for photosynthesis.

7.4 Heterotrophic Protists

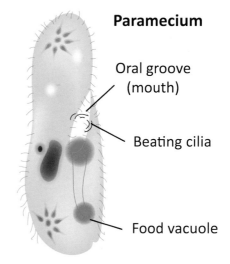

Paramecium

Oral groove (mouth)

Beating cilia

Food vacuole

Many protists do not have the ability to make their own food through photosynthesis. They need to eat, just like we do. Organisms that cannot make their own food are called heterotrophs. Heterotroph comes from the Greek words *hetero* which means "other or different" and *trophe* which means "food or nourishment," so heterotrophs need to find food from sources other than themselves.

Paramecia, for example, live on bacteria, algae, and other small organisms. They have an oral groove that acts just like a big mouth. They gather their food by rapidly beating the cilia near the oral groove and creating water currents that sweep the food into the opening. The food travels into a food vacuole, which is like a tiny stomach for the paramecium. Once food is inside, the vacuole

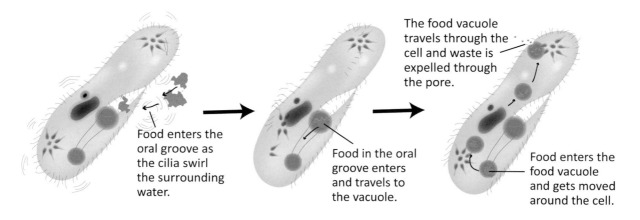

Food enters the oral groove as the cilia swirl the surrounding water.

Food in the oral groove enters and travels to the vacuole.

The food vacuole travels through the cell and waste is expelled through the pore.

Food enters the food vacuole and gets moved around the cell.

circulates around the cell as the food is being digested. Any undigested food left in the food vacuole is ejected through a small pore.

Amoebae are another type of protist that cannot make their own food. Amoebae are hunters; they feed on algae, other protozoa, and even other amoebae. However, an amoeba won't eat everything that comes its way. It is a picky eater.

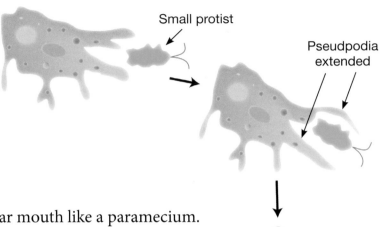

Small protist

Pseudpodia extended

Food vacuole with captured prey

An amoeba does not have a cellular mouth like a paramecium. It can swallow food anywhere on its body. When an amoeba encounters something tasty, it thrusts its pseudopodia outward to surround the prey.

The surrounding membranes then form a food vacuole where the prey is digested. As the food is digested, the food vacuole gets smaller in size as the nutrients are passed into the cytoplasm. Once all of the food has been digested, the food vacuole shrinks and the waste is eliminated through the body surface.

Phagocytosis is the process of eating food by surrounding it and is used by both paramecia and amoebae. Phago comes from the Greek word *phagein* which means "to eat." Cyto comes from the Greek *kytos* which means a receptacle or container. The word-forming element cyt- is used by biologists to refer to a cell, so phagocyte is "a cell that eats."

There are still other protists that use entirely different methods for capturing and consuming food. A didinium, for example, has a single small tentacle called a toxicyst which contains a substance that is poisonous to paramecia. A didinium pierces a paramecium with the toxicyst to paralyze it and then swallows the paramecium

A didinium

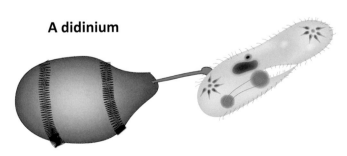

A didinium uses a toxicyst to capture a paramecium

Courtesy of
Gregory Antipa
(San Francisco State University)

whole. Didinia are barrel shaped and have bands of cilia around their body, allowing them to swim fast and move in different directions.

Podophrya, on the other hand, have many tentacles with knobbed ends. A podophrya begins its life as a free-swimming ciliate. When it matures, it loses its cilia, grows tentacles, and uses a stalk to attach itself to a surface. When a protist is swimming past, the podophrya bends and moves to try to capture the prey. If the passing protist touches a tentacle, it sticks to the tentacle and becomes paralyzed. The podophrya

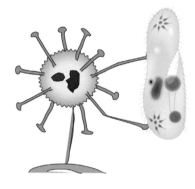

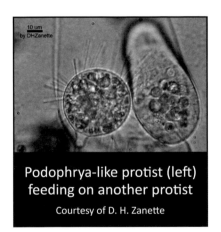

then uses its tentacle to put enzymes into the captured protozoa to break it down into molecules that can be absorbed by the podophrya for food.

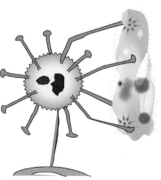

Podophrya-like protist (left) feeding on another protist

Courtesy of D. H. Zanette

Protists are truly remarkable creatures that accomplish an amazing variety of tasks — all within a single cell!

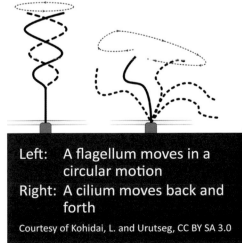

Left: A flagellum moves in a circular motion
Right: A cilium moves back and forth

Courtesy of Kohidai, L. and Urutseg, CC BY SA 3.0

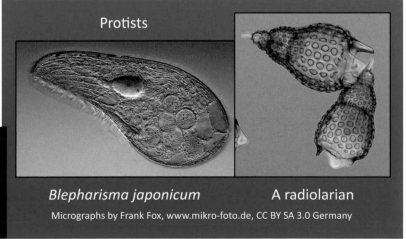

Protists

Blepharisma japonicum   A radiolarian

Micrographs by Frank Fox, www.mikro-foto.de, CC BY SA 3.0 Germany

7.5 Summary

- Protists are microscopic, one-celled organisms that have both plant-like and animal-like qualities.

- There are four main types of protists that are classified primarily on how they move. These are ciliates, flagellates, amoebae, and sporozoans.

- Ciliates move with tiny hair-like projections called cilia.

- Flagellates move with one or more long whip-like structures called flagella.

- Amoebae move by crawling with pseudopodia, or "false feet."

- Photosynthetic protists (autotrophs) , such as euglena, use the Sun's energy to make food.

- Heterotrophic protists, including paramecia, amoebae, didinia, and podophrya capture other organisms for food by using cilia, pseudopods, or tentacles.

7.6 Some Things to Think About

- Study the protist examples in this chapter. What features can you observe that make the organisms different from each other?
  Is it surprising that each of these organisms is made of only one cell (with the exception of the kelp colony) ? Why or why not?

- How would you describe the four ways that protists move that are used for classification?

- How do you think a euglena is helped by having a flagellum and a stigma?

- How would you compare different parts of a protist to different parts of the human body?

- How do you think protists can perform so many different functions when they are made of only one cell?

- What do you think are the advantages of organisms that have many cells compared to organisms that have only one cell?

Image credits for chapter title page:
1. Giant kelp, Courtesy of Claire Fackler, CINMS/ NOAA;   2. Formanifora, Courtesy of Psammophile (CC BY SA 3.0);   3. Star Radiolarian, Courtesy of Dr. John R. Dolan, Laboratoire d'Oceanographique de Villefranche; Observatoire Oceanologique de Villefrance-sur-Mer;
4. Giardia on intestinal cell, Courtesy of CDC/Dr. Stan Erlandsen;   5. Giardia, Courtesy of CDC/Dr. Stan Erlandsen;   6. *Didinium nasutum* eating a paramecium, Courtesy of Gregory Antipa (San Francisco State University) and H. S. Wessenberg (San Francisco State University)

Chapter 8 Fungi: Molds, Yeasts, Mushrooms

8.1 Introduction

What grows in the dirt like a plant but is not green and isn't a plant? Mushrooms! Although mushrooms grow in the dirt like plants and at one time scientists grouped mushrooms with plants, we now know that mushrooms are not plants. Mushrooms are one type of organism in a group of living things called fungi.

Fungi is the plural form of the word fungus which is the Latin word for mushroom. Fungi are in the domain Eukarya and have their own kingdom called Fungi that is separate from plants, animals, and protists. The study of fungi is called mycology, which comes from the Greek word *mykes,* meaning "slippery." If you have ever touched the top of a mushroom or felt the mold on bread, you might remember that it felt slippery.

Photo credits: 1. Steve Hillebrand, USFWS; 2. USFWS; 3. Mold in a petri dish, Scott Bauer, USDA/ARS; 4. Jennifer Winston, USFWS; 5. Public Domain; 6. Chelsi Hornbaker, USFWS

There are around 100,000 known species of fungi that include yeasts, mildews, molds, and mushrooms, and it is thought that there are many more species that have not yet been discovered. Fungi are found in large numbers in many places on Earth including lakes, soils, the air, rivers, and oceans. We use some fungi for food. For example, many different types of non-toxic mushrooms and truffles are eaten in soups, stews, and on pizza! Other fungi, such as yeast and molds, are used in bread, cheese, and beer making.

Fungi come in a variety of different shapes, sizes, and colors. Many of the brilliantly colored mushrooms are beautiful to see but are poisonous if eaten. The smallest fungi, such as yeasts, are microscopic and have only one cell, while other fungi are multicellular. Some fungi are very large. The Phellinus ellipsoideus found in China can form a massive fruiting body (the part of a fungus that produces spores for reproduction). One Phellinus ellipsoideus found on the Hainan Island in southern China, measured 10.8 meters (35.4 ft.) long and almost a meter (3 ft.) wide.

Fungi are not mobile, like animals are, and unlike plants, fungi do not make their own food. Fungi get food from other organisms, some dead and some alive. Many fungi are saprophytes, which means that they use dead or decaying matter for food. Other fungi are parasites and feed on living things. Soils that contain an abundance of organic matter are an excellent environment for fungi, and many species can be found on the forest floor and growing on decaying trees.

8.2 Classification of Fungi

Fungi were originally classified in the plant kingdom, but we now know that fungi don't have chlorophyll and don't produce their own food; therefore, they are very different from plants. The kingdom Fungi is divided into three, four, six or more phyla depending

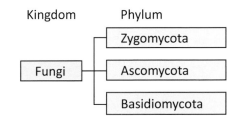

on the way different characteristics are classified. Some scientists divide the kingdom into three phyla based on how fungi reproduce. Other scientists classify fungi based on their molecular biology. In this text we will explore the three phyla Zygomycota, Ascomycota, and Basidiomycota, which are defined by how the fungi reproduce.

1, 2. Grapefruit mold & mold cultures—Courtesy of Scott Bauer, USDA/ARS; 3. Morel; 4. Aleuria Aurantia; 5. Candida yeast—Courtesy of CDC/Melissa Brower (artist's rendition); 6. Chicken of the Woods—Courtesy of National Park Service; 7. Bracket fungi—Courtesy of USFWS/Chelsi Hornbaker; 8 & 9. Mushrooms with caps; 10. Mushroom showing gills—Courtesy of National Park Service

8.3 Structure of Fungi

With the exception of yeasts, the main part of a fungus is called the mycelium. The mycelium is a mass of tiny, thread-like structures called hyphae (singular, hypha) that resemble the roots of plants. Because the hyphae are so numerous, they provide the fungus with a large amount of surface area for absorbing nutrients. Since hyphae can absorb only small molecules, they first release digestive enzymes to break down the organic

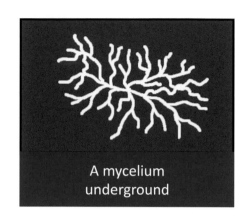

A mycelium underground

matter that surrounds them. The hyphae then absorb the smaller molecules for the fungus to use for food. Each hypha is surrounded by a tough outer cell wall made of chitin, which is the same carbohydrate that insects use to build their exoskeletons.

8.4 Reproduction of Fungi

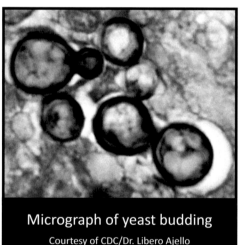

Micrograph of yeast budding

Courtesy of CDC/Dr. Libero Ajello

Fungi can reproduce both asexually and sexually. In asexual reproduction an organism reproduces itself without combining its genetic material with the genetic material of another organism. Asexual reproduction occurs in yeast and some other fungi in the form of budding or fission when one cell divides into two or more cells. All the resulting cells will have the same genetic makeup as the original cell. Another way many fungi can reproduce asexually occurs when a small piece of a hypha breaks off and starts growing into a new organism, or mycelium.

Each of the different fungi phyla presented in this chapter has slightly different ways of reproducing sexually. Sexual reproduction involves the sharing of genetic material between two organisms and is more complicated than asexual reproduction. Fungi do not have male and female organisms. Instead they have different mating types with each mating type containing different genetic material. Some species of fungi have two mating types, while other species have more than two. In general, sexual reproduction in fungi occurs when hyphae of two different mating types touch and exchange genetic material. The hyphal cells then merge the genetic material and make more cells through a series of specific steps in which the merged genetic material is copied. When a new mycelium is formed, it will have the merged genetic material which is different from the genetic material of either of the parent mycelia. This produces a new strain of the fungus. A strain is a group of organisms that have the same genetic makeup as each other.

8.5 Phylum Zygomycota

Zygomycota is the phylum that contains molds. Since moldy food is not safe to eat and tastes bad, we often think of molds as our enemy. But molds, along with other fungi and bacteria, are necessary for breaking down dead plant material and turning it into soil. If this plant material were not broken down, or decomposed, by fungi and bacteria, living plants would not be able to get the nutrients they need to grow.

Some of the organisms in the phylum Zygomycota are among the fungi that form mycorrhizae, which are symbiotic relationships between fungi in the soil and plant roots. A symbiotic relationship is one that is beneficial to both organisms. In a mycorrhiza, the fungus growing on a plant's roots sends out hyphae that spread farther out into the soil. Some of he nutrients the fungus absorbs become available to the plant, giving the plant access to nutrients that are beyond the reach of the plant's root system. The amount of nutrients available to the plant increases, and the plant in turn provides the fungus with nutrients that the fungus needs to grow and function, such as sugars that the plant makes through photosynthesis.

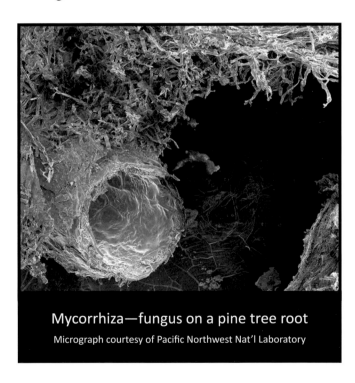

Mycorrhiza—fungus on a pine tree root
Micrograph courtesy of Pacific Northwest Nat'l Laboratory

The zygomycetes (members of the phylum Zygomycota) have different mating types for sexual reproduction. After the hyphae of two zygomycetes of different mating types touch, they form a bridge between them. Genetic material from each of the hyphae goes into the bridge and is merged. The bridge with the merged genetic material grows into a new cell called a zygosporangium. The zygosporangium is a spore from which a new mycelium can grow. Since the zygosporagium contains merged genetic material from two different mating types, the new organism will have different genetic material than either of the original mating types.

When conditions are unfavorable for growth, the zygosporangium rests, or doesn't grow. The zygosporangium has a tough outer coat to protect it and is able to survive harsh conditions such

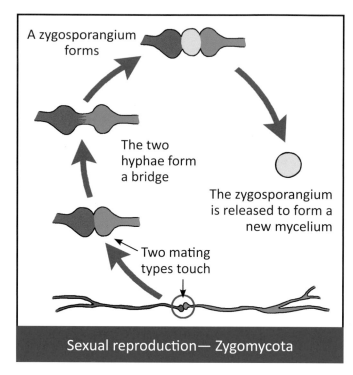

A zygosporangium forms

The two hyphae form a bridge

The zygosporangium is released to form a new mycelium

Two mating types touch

Sexual reproduction— Zygomycota

as freezing or drying. When conditions improve, the zygosporangium can germinate and become a new organism. The phylum Zygomycota is named for the tough zygosporangium.

Zygomycetes can make spores asexually (without having the two mating types come together to share genetic material) by forming a structure called a sporangium where spores are produced. A sporangium forms at the tip of a hypha and looks like a ball mounted on a stem — something like a tree from a Dr. Seuss book! When the spores are mature, they are released from the sporangium. All of the spores will have the same genetic makeup.

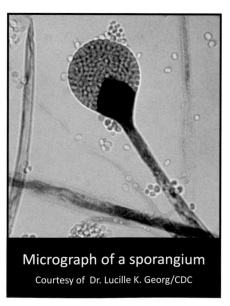

Micrograph of a sporangium

Courtesy of Dr. Lucille K. Georg/CDC

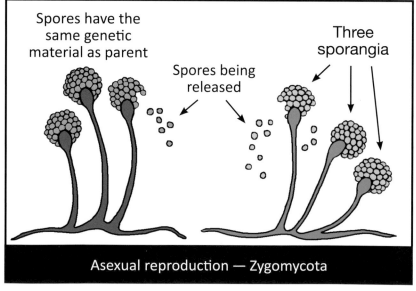

Spores have the same genetic material as parent

Spores being released

Three sporangia

Asexual reproduction — Zygomycota

8.6 Phylum Ascomycota

The phylum Ascomycota is named for a reproductive feature called an ascus. During sexual reproduction, a fungus in the phylum Ascomycota produces asci (plural form). Each ascus is a sac that contains microscopic spores called ascospores. When an ascospore is released by the fungus and lands in a location that has the right conditions, a new organism will grow.

Like zygomycetes, the ascomycetes have different mating types. When two hyphae of different mating types touch, they join and form a bridge. The genetic material of one hypha moves into the other hypha. Next, the hypha containing the genetic material from both mating types grows to make more hyphae that will create an ascocarp, the fruiting body of the fungus. When the ascocarp has formed, there will be tips (ends) of hyphae on the ascocarp's inner surface. Asci form on these hyphal tips, and ascospores are produced in the asci. When the ascospores are ready, they will be released from the ascus to grow new organisms that will have genetic material that is different from that of either of the original mating types.

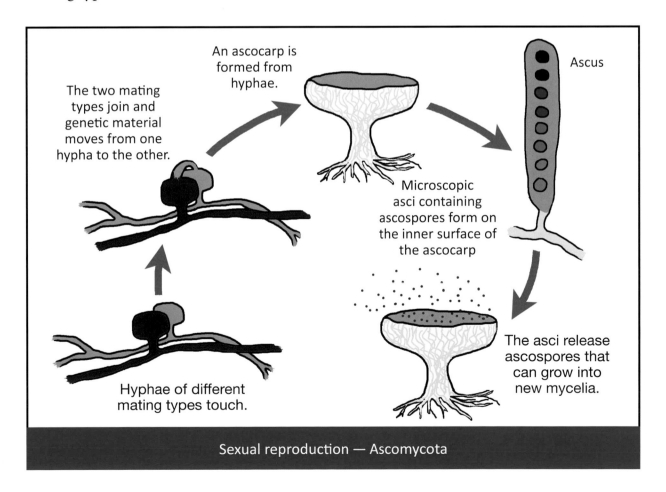

The two mating types join and genetic material moves from one hypha to the other.

An ascocarp is formed from hyphae.

Ascus

Microscopic asci containing ascospores form on the inner surface of the ascocarp

Hyphae of different mating types touch.

The asci release ascospores that can grow into new mycelia.

Sexual reproduction — Ascomycota

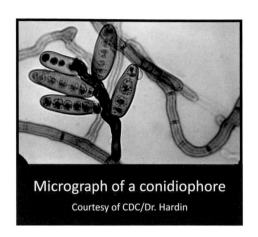

Micrograph of a conidiophore
Courtesy of CDC/Dr. Hardin

Ascomycetes can reproduce asexually by making spores called conidia. The conidia are grouped in structures called conidiophores that grow from the tips of hyphae.

One of the largest of the ascomycetes is the truffle. Truffles are a delicacy in many parts of Europe and North America and are used in soups, stews, and sauces. Truffles can be as small as a pea or as big as an orange. Truffles can be difficult to find because they live as much as 0.3 meter (1 foot) below the surface of the soil. Truffles emit a very strong odor, and truffle hunters are often led to the truffles by trained female pigs that are sensitive to the scent. The truffle that we eat is the ascocarp of the organism.

The yeasts we use in food are among the smallest and most familiar members of the phylum Ascomycota. Yeasts are microscopic and reproduce asexually by budding. Under extreme conditions yeasts can reproduce sexually by producing ascospores, although the process is slightly different from that described previously.

Yeasts are used to make many foods, such as beer, wine, and bread. When yeast is added to sweetened bread dough, the yeast uses the sugar to make carbon dioxide and ethanol (alcohol). The bubbles of carbon dioxide make the bread dough rise and leave air holes in the baked bread. The alcohol that is being created gives raw bread dough a tangy or sour taste, but the alcohol doesn't stick around for long. When the bread dough is cooked, the ethanol is evaporated by the heat of the oven.

8.7 Phylum Basidiomycota

The phylum Basidiomycota is named for its reproductive structure which defines the shape of mushrooms. The name Basidiomycota comes from the Latin word *basidium* which means "small pedestal." There are several thousand species of fungi in the phylum Basidiomycota, and these include mushrooms, shelf fungi, puffballs, and rusts.

Mushrooms showing gills
on the bottom of the caps

Like other fungi, basidiomycetes can reproduce asexually or sexually. Asexual reproduction occurs when a hyphal cell goes through the budding process. Recall that during budding, a cell divides in two. The new cell formed by this process separates from the parent cell and has the same genetic material as the parent cell. The new cell can grow into a new organism when the conditions are right. Another type of asexual reproduction in basidiomycetes occurs at the tip of a hypha where a structure (conidiophore) forms to make spores (conidia) that can be released to grow into new organisms.

Basidiomycetes have different mating types that come together during sexual reproduction. When two hyphae that have different mating types touch, they exchange genetic material and make spores that can grow into new mycelia.

In mushrooms, when the underground mycelium is ready to reproduce, a hypha that contains genetic material of different mating types will grow into a knot. Hyphae will then grow from the knot into an above ground mushroom (the fruiting body) that has a stem with a cap, or umbrella, on top. The hyphae that make up the cap grow downward, forming thin ribs called gills. Spores are produced at the ends of hyphae in the gills on the underside of the cap. The spore-producing

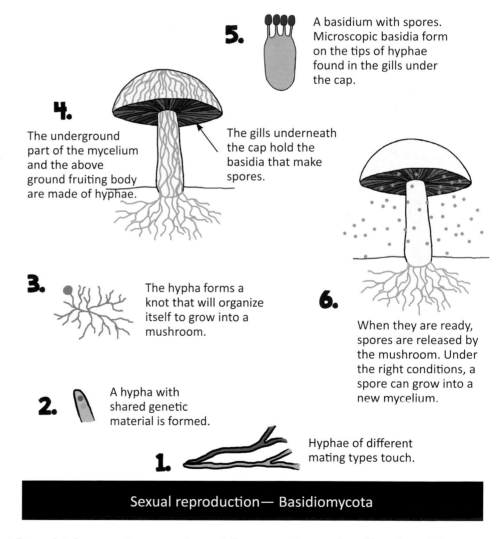

5. A basidium with spores. Microscopic basidia form on the tips of hyphae found in the gills under the cap.

4. The underground part of the mycelium and the above ground fruiting body are made of hyphae.

The gills underneath the cap hold the basidia that make spores.

3. The hypha forms a knot that will organize itself to grow into a mushroom.

6. When they are ready, spores are released by the mushroom. Under the right conditions, a spore can grow into a new mycelium.

2. A hypha with shared genetic material is formed.

1. Hyphae of different mating types touch.

Sexual reproduction— Basidiomycota

structures are called basidia, which are microscopic and form on the ends of hyphae. The spores made in the basidia are called basidiospores. When they are ready, the basidiospores are released by the mushroom and scattered by the wind. If a spore lands in a spot that has the right conditions, it will grow to make a new mycelium. Because there is such a small chance of a spore landing in a favorable spot, a mushroom will make millions or billions of spores to improve its chances of reproduction.

In the right places, you can sometimes find a perfect circle of mushrooms. These circles have the nickname "fairy rings," and the legends say that the fairies danced in a circle there the night before. Can you guess why the mushrooms grow in a circle?

It turns out that a fungal mycelium can live for many years. After a time of growth, as the mycelium gets bigger, the center dies, and the living mycelium becomes a donut shape, sending up mushrooms along its edge. This creates a "fairy ring" of mushrooms.

8.8 Summary

- Fungi are in a kingdom separate from plants, animals, and protists.

- The study of fungi is called mycology.

- With the exception of yeasts, the main part of a fungus is called the mycelium. The mycelium is a mass of tiny, thread-like structures called hyphae that resemble the roots of plants.

- The phylum Zygomycota includes the molds.

- The phylum Ascomycota includes the yeasts.

- The phylum Basidiomycota includes the mushrooms.

8.9 Some Things to Think About

- What are some fungi you have observed? What features did they have that were different from plants?

- Why do you think some scientists classify fungi based on how they reproduce, while other scientists use molecular biology as a basis for classification?

- In your own words, describe the main part of a fungus.

- If you compare asexual reproduction and sexual reproduction in fungi, do you think one process is more complicated than the other? Why?

- What do you think are the advantages and disadvantages of each type of reproductive process?

- Briefly describe sexual and asexual reproduction in zygomycetes.

- Do you think there are conditions under which a fungus would be more likely to use one reproductive process than the other?

- In what ways do you think sexual reproduction in ascomycetes is similar to that of zygomycetes?

- The next time you're out walking and see a mushroom or you find one on your plate, what facts will you know about it?

Chapter 9 Plants

9.1 Introduction

Everywhere you go there are plants. Plants are one of the most adaptable and abundant forms of multicellular life on the planet. In many parts of the world plants mark the coming and going of the seasons as spring flowers blossom after cold winters, and vegetables, fruits, and tubers are harvested in the fall. Plants produce food for the world's population and color the landscapes with rolling hills of grass and vast forests of trees.

Plants belong to the domain Eukarya and the kingdom Plantae. Plants include all the trees, grasses, flowering plants, shrubs, mosses, and ferns.

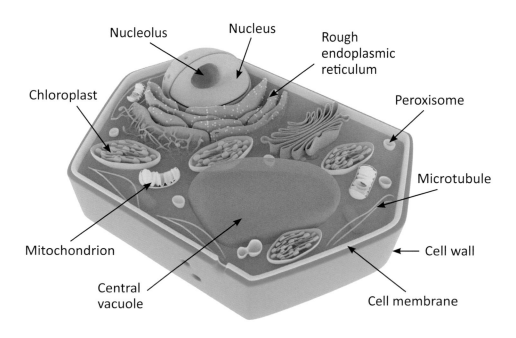

Photo credits: 1. *Galeopsis speciosa*, by André Karwath, CC BY-SA 2.5; 2. Fern, by Piotr Menducki; 3. Orange tree, by Jose Luis Navarro

9.2 Plant Cells

Nucleolus Nucleus Rough endoplasmic reticulum

Chloroplast

Peroxisome

Microtubule

Mitochondrion

Cell wall

Central vacuole

Cell membrane

Plants are made of eukaryotic cells that contain organelles. Organelles are small structures that act like "tiny organs" inside the cell and perform different functions, like making energy molecules and manufacturing proteins. Many organelles are surrounded by a cell membrane. The nucleus is an organelle in eukaryotic cells that holds DNA and other protein machinery needed to copy the genetic material used for reproduction. The organelles found in plant cells that are similar to those found in animal cells include mitochondria which are used to make energy molecules, the rough endoplasmic reticulum where proteins are made, and peroxisomes that help convert fatty acids to sugars.

Plant cells also contain organelles called plastids. Plastids are surrounded by a double cell membrane and are small specialized organelles that perform functions specific to plants. Chloroplasts are a type of plastid used to convert the Sun's energy into food through the process of photosynthesis. Chloroplasts contain chlorophyll, which is a molecule called a pigment that gives plants their green color. Other plastids include leucoplasts, which store a variety of energy sources, such as starches, and

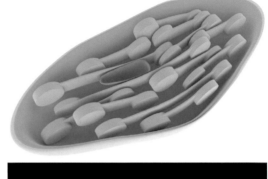

A chloroplast

chromoplasts, which lack chlorophyll but have other molecules called carotenoids that give plants, flowers, and fruit their yellow, orange, and red colors. Another important organelle in plant cells is the central vacuole which stores chemicals and breaks down proteins.

Like animal cells, each plant cell is surrounded by a cell membrane, but unlike animal cells, a plant cell also has a cell wall. The cell wall is a protective layer outside the cell membrane. Cells walls are made of cellulose, a linked glucose polymer chain that gives plants the support they need to stand tall and hold the shape of their leaves.

Molecular structure of cellulose

9.3 Classification of Plants

Like other organisms, Latin names are used for the taxonomic classification of plants. Plants are in the domain Eukarya and the kingdom Plantae. The kingdom is further divided into phylum (or division[1]), class, order, family, genus, and species.

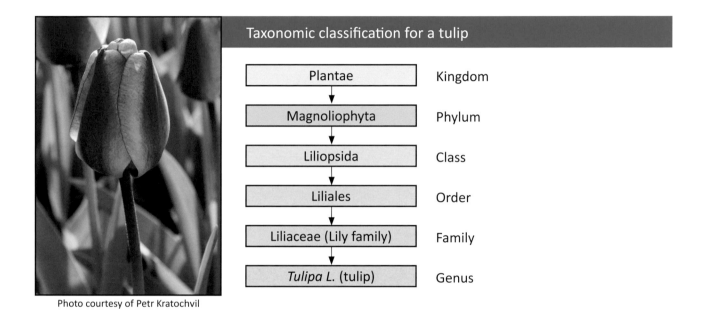

Photo courtesy of Petr Kratochvil

Taxonomic classification for a tulip

Plantae	Kingdom
Magnoliophyta	Phylum
Liliopsida	Class
Liliales	Order
Liliaceae (Lily family)	Family
Tulipa L. (tulip)	Genus

Botanists group plants into different phyla within the kingdom Plantae according to the plants' common shared features. Sorting plants into any group is challenging because of their diversity, and as a result, botanists do not always agree on the number of phyla to use when grouping plants. Some sources cite close to 30 phyla, while others cite between 7 and 12. In this text we will divide plants into eleven phyla which cover all of the present day plants known at this time.

Although there are many different ways to group plants, there are two broad generic categories that separate plants based on how water and nutrients travel through plant tissue. These two groups are called vascular and nonvascular. We will learn more about nonvascular plants later in this chapter.

---

1. The term "division" has traditionally been used when sorting plants below the kingdom level and above the class level. However, in 1993 the International Code of Nomenclature for algae, fungi, and plants accepted the designation "phylum" for this level. Many references still use division, and it means the same thing as phylum.

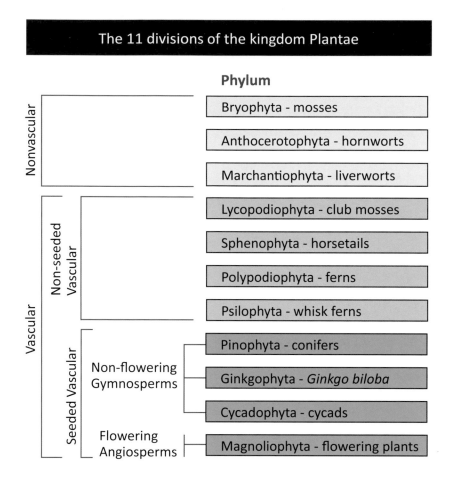

9.4 Nonvascular Plants

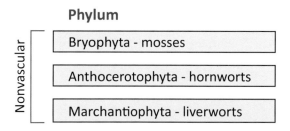

Nonvascular plants do not have xylem and phloem, the vascular tissues that are used by vascular plants to carry water and nutrients through the plant. Instead, nonvascular plants absorb water and nutrients from their surroundings through their leaf surfaces or plant body. Because nonvascular plants lack the structural fibers found in xylem and phloem, they don't have the structural tissues to stabilize growth and are therefore small and low to the ground. They need to live in moist environments where their plant body can be in contact with water.

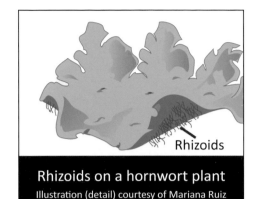

Rhizoids

Rhizoids on a hornwort plant
Illustration (detail) courtesy of Mariana Ruiz

Nonvascular plants lack roots and instead have structures called rhizoids—small hairs that dig into the soil to keep the plant in place. Rhizoids are not classified as true roots because they do not contain vascular tissue.

Nonvascular plants also have simple reproductive methods. They can reproduce sexually by creating single-celled spores, and they can reproduce asexually by vegetative propagation. Vegetative propagation occurs when part of the plant breaks off and develops into a new plant that has genetics that are identical to the original plant.

Phylum Bryophyta

Mosses are the most common nonvascular plants with over 14,000 known species. Mosses can be found in bogs and swamps at sea level and also at high altitudes on mountains. Although some mosses can be found in deserts, most are found in moist, shaded habitats. Mosses are small and low to the ground, can have a variety of colors, and can grow up trees, across fields, and on top of rocks.

Moss on a wall and close up of spore producing structures
Unidentified moss on rock wall, Fred Hsu, CC BY-SA 3.0; Close up of spore producing structures of unidentified moss, by Rostislav Kralik

Hornwort
Courtesy of Jason Hollinger, Wikimedia Commons CC BY-SA 2.0

Phylum Anthocerotophyta

Hornworts are nonvascular plants found in ponds, in slow streams, on riverbanks, and in other damp places. Because hornworts don't have roots, they use their leaves to take in nutrients directly from the water. Hornworts that grow submerged in water can form large, branched masses. These may be attached to the bottom of a pond or stream by rhizoids, or they can float around in the water. The plant bodies of hornworts are green and flattened with a waxy appearance. Hornworts are named for their slender hornlike structures. The "wort" part of the name comes from the Old English *wyrt* which means "herb" or "plant."

Phylum Marchantiophyta

Liverworts are nonvascular plants similar to hornworts but lacking hornlike structures. There are over 10,000 known species of liverworts. They got their name because at one time it was believed that liverworts could be used to cure diseases of the liver. Most liverworts are small and some have leaves. Others have a plant body called a thallus. Like hornworts and mosses, liverworts live in moist environments.

Liverwort

Top: Courtesy of Hermann Schachner Right: Courtesy of Frank Vincentz, Wikimedia Commons, CC BY SA 3.0

9.5 Vascular Plants

Vascular plants include all plants that contain the structures called xylem and phloem, the vascular tissues that are used to carry water and nutrients through the plant. The word vascular means having "vessels" or "tubes." We will find out more about vascular tissues in Chapter 11. Vascular plants are divided into two broad groups: seeded

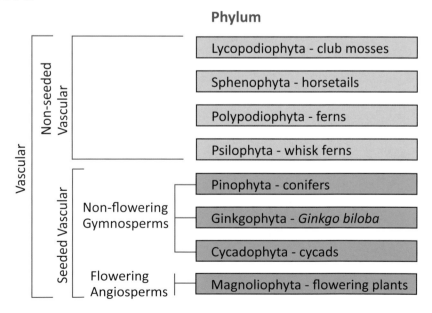

Phylum

Vascular

Non-seeded Vascular

Seeded Vascular

Non-flowering Gymnosperms

Flowering Angiosperms

Lycopodiophyta - club mosses

Sphenophyta - horsetails

Polypodiophyta - ferns

Psilophyta - whisk ferns

Pinophyta - conifers

Ginkgophyta - *Ginkgo biloba*

Cycadophyta - cycads

Magnoliophyta - flowering plants

vascular plants and non-seeded vascular plants. Seeded vascular plants are divided further into flowering and non-flowering plants.

9.6 Non-seeded Vascular Plants

Like the description implies, non-seeded vascular plants don't reproduce using seeds. Although seedless vascular plants can grow in many different habitats, they require some moisture or open water, such as lakes or ponds, to reproduce and grow. Non-seeded vascular plants include club mosses, ferns, and horsetails.

Phylum

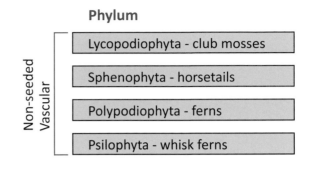

Non-seeded Vascular

| Lycopodiophyta - club mosses |
| Sphenophyta - horsetails |
| Polypodiophyta - ferns |
| Psilophyta - whisk ferns |

Club mosses
Photo credits (Wikimedia Commons): 1. *Diphasiastrum digitatum,* by Jaknouse, CC BY-SA 3.0; 2. *Lycopodium annotinum,* by Alastair Rae, CC BY-SA 2.0

Phylum Lycopodiophyta

Club mosses are small vascular plants in the family Lycopodiaceae. Club mosses live near streambeds and marshes and in moist woodlands and have needlelike or scalelike leaves.

Phylum Sphenophyta

Horsetails are in the family Equisetaceae and are strange-looking vascular plants with jointed, upright, hollow stems and creeping horizontal stems called rhizomes which grow several feet deep into the ground. Horsetails have no true leaves and may have a bare stem, or they may have whorls of small branches around the stem.

Horsetail
Photo credits (Wikimedia Commons): 1. *Equisetum telmateia,* by Rorr, CC BY SA 3.0; 2. *Equisetum hyemale,* by Line1, CC BY SA 3.0

Phylum Polypodiophyta

With over 11,000 species, ferns are the largest group of non-seeded vascular plants. Ferns are found in a wide variety of areas including deserts, coastal areas, and forested lands. The leaf of a fern is called a frond and is divided into a stipe, which is the leaf stalk, and the blade, the leafy portion. Ferns vary in size with some gaining a height of over 10 meters. Fronds can be longer than 3 meters. Tiny ferns may have fronds that are only about 2 cm long.

Ferns
Courtesy of Marc Ryckaert, Wikimedia Commons, CC BY-SA 4-0

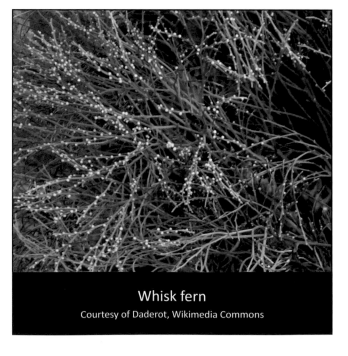

Whisk fern
Courtesy of Daderot, Wikimedia Commons

Phylum Psilophyta

Whisk ferns are a small group of vascular plants found in tropical and subtropical regions. Whisk ferns have stems that grow upward to a height of about .3-.6 meters and rhizomes that grow underground. They don't have leaves, and their spores are formed in yellow spherical structures on the stems. The name whisk fern comes from a time in the past when people would tie branches of these plants together to make a whisk broom.

9.7 Seeded Vascular Plants

Seeded vascular plants are the largest group of plants in the plant kingdom. If you've ever eaten an ear of corn, pine nuts, or pecans, you have eaten the seeds of a vascular plant. A plant seed contains the embryo, which is the beginning stage of a plant and will

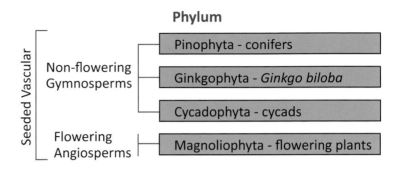

grow into a mature plant. The nutrients that the embryonic plant needs to grow are also contained within the seed. The seed is covered in a tough case which helps it to survive until conditions are right for it to begin to grow. If a seed is placed in a suitable environment, a small seedling will sprout from the seed. The seedling will have fully functioning tissues in its plant body that form roots and perform photosynthesis. Because the seed is covered in a tough case, seeds allow plants to grow with less water than the amount needed by non-seeded plants. Also, the seed contains food for the embryo to use as it begins to grow.

Seeded vascular plants are divided into two groups: non-flowering, cone-bearing plants called gymnosperms, and flowering plants called angiosperms. Gymnosperms include conifers, cycads, and ginkgos. Angiosperms include all the grasses, flowering trees and shrubs, and all the common flowering plants such as sunflowers, orchids, and roses.

Gymnosperms

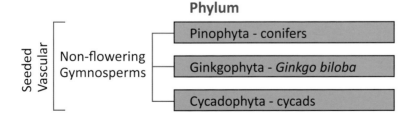

Gymnosperms have seed-bearing cones and do not produce flowers. The word gymnosperm means "naked seed." Unlike flowering plants, gymnosperm seeds are not covered with fruit.

Phylum Pinophyta

Conifers are the largest division of the gymnosperms. Many are found in mountainous regions, and many do well in cold, dry climates. To reduce water loss, these trees and shrubs have thin needlelike leaves covered in a waxy, waterproof layer called the cuticle. Many

Conifers at Mount Rainier, Washington State

Courtesy of Alan Cressler (alan cressler on Flickr)/USGS

conifers grow in a cone shape with a wide base, narrow top, and branches that point downward, allowing them to shed snow in the winter months without breaking their branches. The majority of conifer forests occur in the Northern Hemisphere, and some conifers grow in warmer climates. In general, conifer trees produce soft wood in the stem, or trunk, of the tree, and conifers are used to make paper and construction materials.

Phylum Ginkgophyta

Ginkgo biloba is unique among the gymnosperms because it is the only plant in its division—it has no other living relatives. Fossil records show that ginkgo trees existed over 200 million years ago, and the ginkgo of today is thought to be almost identical to the fossilized trees. *Ginkgo biloba* trees can grow to be over 24 meters high with a spread of 12 or more meters. They have fan-shaped green leaves that turn a golden color in the autumn when the leaves are ready to be shed.

The seeds of the ginkgo have been used for medicinal purposes for many centuries and are believed to improve conditions such as short-term memory loss, inflammation, and fungus infections. *Ginkgo biloba* trees are widely cultivated, but there are few wild populations.

Ginkgo biloba tree and leaf

Credits: Ginkgo tree, User: Sunroofguy, Wikimedia Commons, License CC BY SA 3-0; Ginkgo leaf, Davide Guglielmo

Phylum Cycadophyta

Cycads are slow growing trees that have very long life spans, with some specimens as old as 1000 years. According to fossil records, cycads were plentiful more than 200 million years ago, and like ginkgos, they have not changed much since then. There are about 300 species of cycads currently existing, and many more have gone extinct due to habitat loss, infrequent reproduction, and slow growth. Most cycads live in tropical or subtropical environments and vary in size from a few centimeters to about 18 meters high. Cycads generally have a thick, woody trunk topped by a crown of large leaves.

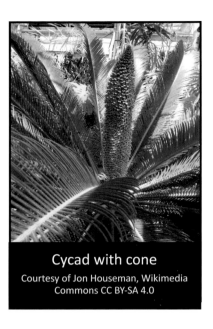
Cycad with cone
Courtesy of Jon Houseman, Wikimedia Commons CC BY-SA 4.0

Angiosperms

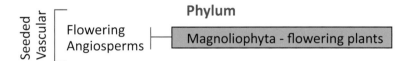

Angiosperms, or flowering plants, are the most common plants on Earth with over 400,000 different species. Angiosperms have flowers and fruit and seeds that are covered by a shell. Angiosperms use flowers to recruit bees and other small animals to do the work of pollination—spreading the genetic material contained in the pollen from plant to plant. We will learn more about pollination and reproduction of flowering plants in Chapter 12.

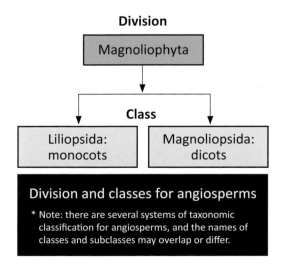

Angiosperms have traditionally been divided into two main classes—Liliopsida, or monocots, and Magnoliopsida, or dicots. These groupings have not always been agreed upon by botanists and continue to be challenged as new research brings new information about the details of these two types of flowering plants. However, there are distinct differences, and separating them into two groups is still useful even though there are some exceptions, with some plants having characteristics that fit into both groups.

Monocots

Corn seed

Cotyledon

Embryo

Embryos have a single cotyledon

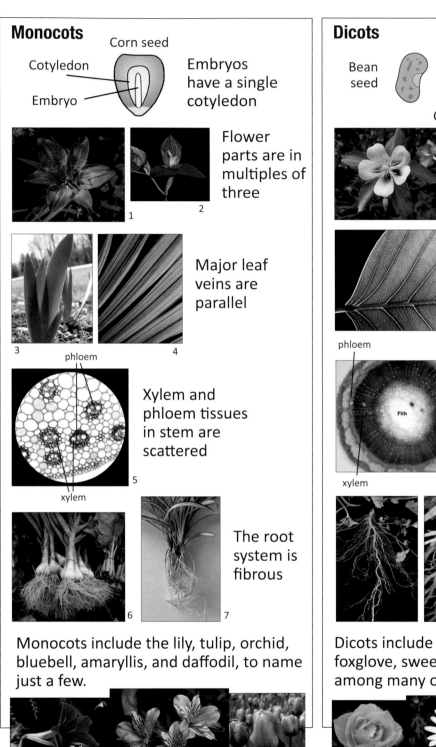

Flower parts are in multiples of three

1

2

Major leaf veins are parallel

3

4

phloem

Xylem and phloem tissues in stem are scattered

5

xylem

The root system is fibrous

6

7

Monocots include the lily, tulip, orchid, bluebell, amaryllis, and daffodil, to name just a few.

13

14

15

16

17

18

19

Dicots

Embryo

Bean seed

Embryos have two cotyledons

Cotyledons

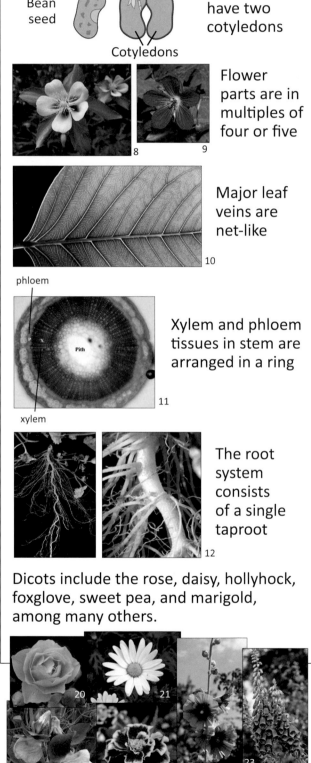

Flower parts are in multiples of four or five

8

9

Major leaf veins are net-like

10

phloem

Pith

Xylem and phloem tissues in stem are arranged in a ring

11

xylem

The root system consists of a single taproot

12

Dicots include the rose, daisy, hollyhock, foxglove, sweet pea, and marigold, among many others.

20

21

24

25

22

23

Photo credits at end of chapter.

The terms monocot and dicot are short for monocotyledon and dicotyledon. *Mono* means one, so monocots have one cotyledon. *Di* means two, so dicots have two cotyledons. When a monocot seed begins to grow, or germinates, the cotyledon is the part of the seed that develops into the single first leaf. In a dicot seed, the cotyledon is the part of the seed where food is stored for the germinating plant. A dicot seed has two cotyledons, and the first leaves will be a set of two.

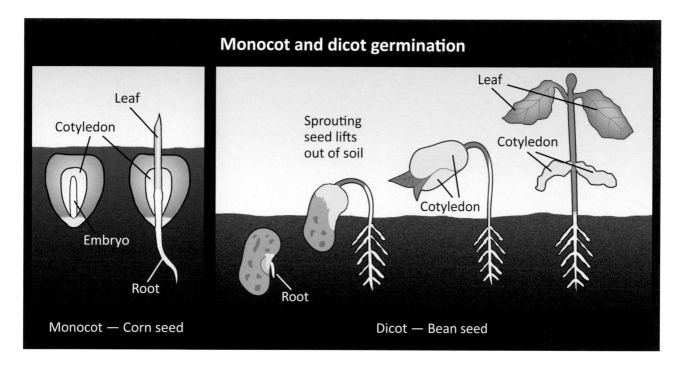

9.8 Summary

- Plants belong to the domain Eukarya and the kingdom Plantae.

- Like animal cells, plant cells contain organelles, such as mitochondria and the nucleus. Plant cells also contain plant-specific organelles called plastids which include chloroplasts.

- The kingdom Plantae is divided into different phyla.

- There are two main types of plants: nonvascular plants and vascular plants. Nonvascular plants do not contain vascular tissues.

- Vascular plants are sorted into two main divisions: flowering plants and non-flowering plants.

9.9 Some Things to Think About

● In what ways do you think plants are important to your life and to life on Earth?

● Describe a plastid.
Name some plastids and describe what each does.

● What is a nonvascular plant?
What are the three phyla of nonvascular plants?

● What is a vascular plant?
What advantages do you think a vascular plant has over a nonvascular plant?
What are the characteristics of a non-seeded vascular plant?

● What are some interesting characteristics of ferns and horsetails?

● Why do you think seeded vascular plants are the largest group of plants?

● Describe the differences between gymnosperms and angiosperms.

● List some differences between monocots and dicots. Do you think you could tell monocots and dicots apart by examining the plants and their flowers?

Chapter 10 Photosynthesis

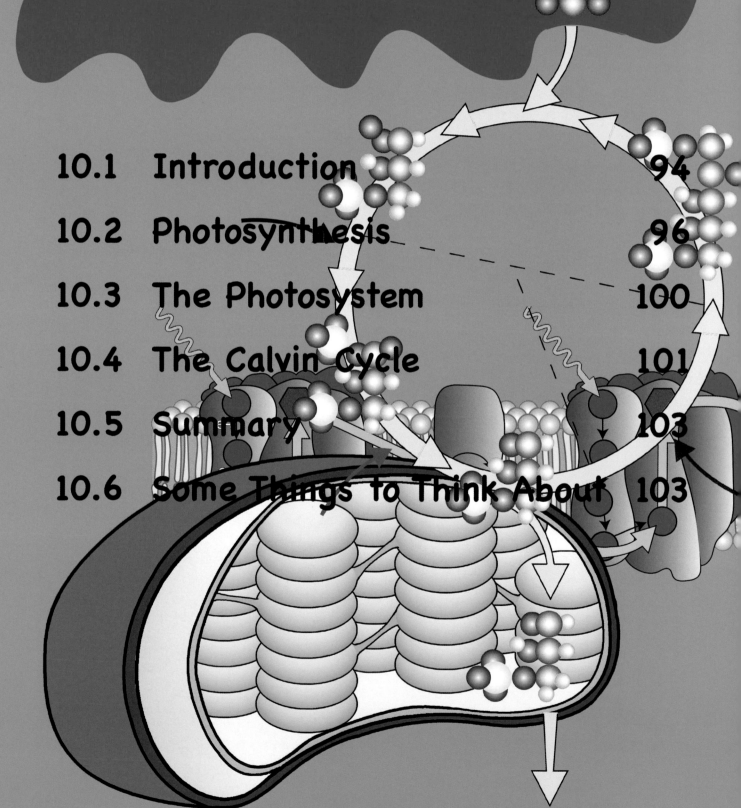

10.1 Introduction

Grasslands in Montana

Courtesy of Dr. David Goodrich, NOAA (ret.)

Plants have the remarkable ability to make food out of sunlight, carbon dioxide, and water. The entire animal kingdom relies upon plants for food— without plants animal life on Earth would cease to exist.

Plants are called autotrophs because they make their own food by a process called photosynthesis. Photo comes from the Greek word *photos*, which means "light" and synthesis, which means "to make." Photosynthesis means "to make with light."

The process of photosynthesis occurs in plastids called chloroplasts. Recall from Chapter 9 that plastids are specialized organelles found in plants and some other organisms. Chloroplasts are bound by both an inner and an outer membrane that allow all the necessary chemical reactions to be carried out in a confined container. Inside the chloroplast are membrane-bound structures called thylakoids that are arranged in stacks called grana (singular, granum).

Within the thylakoid membrane are chlorophyll molecules that give plants their green color. Chlorophyll molecules are carefully arranged in the thylakoid so that when sunlight interacts with one chlorophyll molecule, it starts a cascade of interconnected chemical reactions. Surrounding the thylakoids is the stroma, a colorless liquid that contains supportive fibers that hold the chloroplast together.

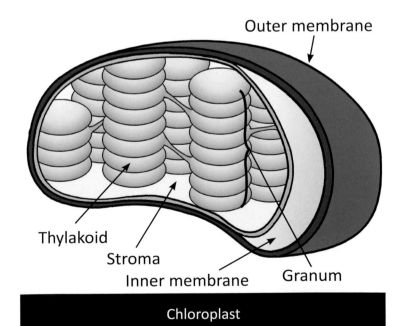

Outer membrane

Thylakoid

Stroma

Inner membrane

Granum

Chloroplast

Chloroplasts are found in all the green parts of a plant including the leaves, stems, and unripened fruit. However, green leaves are the major site of photosynthesis in most plants. Because of their shape and location on the plant, leaves can collect much more sunlight than any other part of the plant. For example, the leaves of flowering plants are broad with a flat shape. This shape enables the leaves to collect as much sunlight as possible. On trees, the leaves are fixed to the branches at many different angles. This enables the leaves to collect sunlight from all directions during the day as the Sun moves across the sky.

Photosynthesis occurs year-round in some plants, for example, conifer trees. Instead of broad, flat leaves, many conifers have needles, very narrow leaves covered by a cuticle—a waxy, waterproof outer coating. Trees that have needles are uniquely designed to live in places where it is drier and where broad, flat leaves would tend to dry out. Because they keep their needles year-round, conifers can carry out photosynthesis during the

Broad leaves and needles
Photo credits: Leaves, Wayne Ray, Wikimedia Commons
License CC BY-SA 4.0; Needles, Petr Kratochvil

winter, and the shape and arrangement of needles help the trees shed snow so the branches don't break from the weight.

Some organisms that are not classified as plants also use the Sun's energy to make food. For example, many types of algae are photosynthetic. Algae are in the kingdom Protista and can be microscopic or macroscopic (large enough to see with the unaided eye). Seaweeds found in the ocean are classified as algae rather than true plants. Although seaweeds have structures that look like those of plants, they are not classified as plants because they lack true roots, stems, and leaves. Seaweeds don't have vascular tissues, and they absorb nutrients directly from the water they are submerged in. Not all seaweeds are green. Some are red or brown. However, all seaweeds do use photosynthesis to make their own food just like land plants do. Microscopic algae make most of the food for the animals in the ocean.

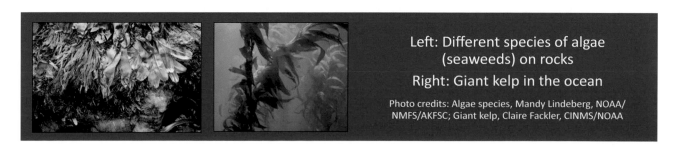

Left: Different species of algae (seaweeds) on rocks
Right: Giant kelp in the ocean

Photo credits: Algae species, Mandy Lindeberg, NOAA/
NMFS/AKFSC; Giant kelp, Claire Fackler, CINMS/NOAA

Another class of organisms that use the Sun's energy to make food are the cyanobacteria. Cyanobacteria were once called blue-green algae, but because the cells don't have a nucleus (they are prokaryotes), they are not like other algae. They are now grouped in the kingdom Bacteria.

10.2 Photosynthesis

Although the exact series of chemical reactions for photosynthesis is complex, the overall photosynthetic reaction has been known since the early 1800s. In summary, light energy, water, and carbon dioxide produce glucose (a simple sugar) and oxygen.

The overall chemical reaction can be written as:

$$6CO_2 + 12H_2O + \text{Light energy} = C_6H_{12}O_6 + 6O_2 + 6H_2O$$

(6 molecules of carbon dioxide + 12 molecules of water + light energy = one molecule of glucose + 6 molecules of oxygen + 6 molecules of water)

The investigation of photosynthesis grew out of a simple question: "When a seedling starts to grow, where does the increase in mass come from?" Does it come from the water? Does it come from the soil? Does it come from the air?

Julius Robert von Mayer
(1814-1878 C.E.)

Beginning in the mid 1600s scientists began exploring the chemistry behind how plants grow to determine where in the process plants create the additional tissue needed to grow into tall trees, long grasses, and bountiful vegetables. In 1845 Julius Robert von Mayer (1814-1878 C.E.) proposed that plants convert light energy into chemical energy, and by the mid 1900s several scientists showed that atmospheric carbon dioxide is turned into simple sugars while releasing oxygen.

The simple photosynthetic equation above is actually a series of two distinct stages of multiple reactions known as the light-dependent reactions and the light-independent reactions, or Calvin Cycle. The light-dependent reactions are the chemical reactions that occur when a chloroplast captures light energy from the Sun, and the Calvin Cycle (historically called the dark reactions) refers to the chemical reactions that produce sugar. One way to look at these two sets of reactions is that the *light-dependent reactions* are the *photo* part of photosynthesis and the *Calvin Cycle* is the *synthesis* part of photosynthesis. We'll take a look at the light-dependent reactions and the Calvin Cycle in more detail later in this chapter.

How does the process of photosynthesis work? How does a plant use sunlight to make food? Recall that light is composed of both electric and magnetic energy that combine to make an electromagnetic wave. Electromagnetic waves have different frequencies and wavelengths. For example, radio waves have low frequencies and long wavelengths, and gamma rays have high frequencies and short wavelengths. The visible light part of the electromagnetic spectrum is a narrow range of wavelengths that are visible to the eye. Visible light wavelengths are between the longer infrared and the shorter ultraviolet wavelengths, both of which require instruments to be observed.

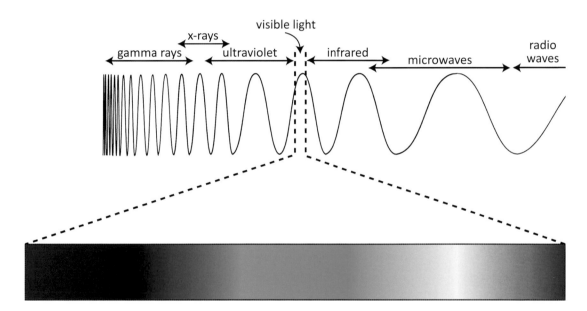

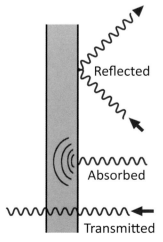

Recall that light can be reflected, absorbed, or transmitted. When a light wave is reflected, it bounces back away from the material. When a light wave is absorbed, energy from the light is transferred to the material that is absorbing it, and when a light wave is transmitted, it passes through the material. The color the material appears to us is the wavelength of visible light that is reflected the most. Plant leaves look green because green light is reflected. This means that violet, blue, red, orange, and yellow light wavelengths are absorbed.

When sunlight hits a plant cell that contains a chloroplast, light is absorbed by special molecules known as pigments. Different pigments absorb and reflect light at different wavelengths depending on the chemical structure of the pigment molecules. Pigments are large molecules with several rings and alternating double bonds. Because they have rings and double bonds, pigments can absorb light of different wavelengths by converting light energy into chemical potential energy inside these rings and double bonds. We'll take a closer look at pigment molecules later, but first let's explore how they can absorb and release light energy. How does this happen?

Recall that light energy is made of photons which are packets of energy that travel through space like a wave. In a nutshell, the way a pigment molecule converts light energy into chemical energy is to take energy from a photon and use this energy to bump one or more of its electrons into a higher energy state. Most atoms and molecules exist in what is called the ground state. The ground state is the lowest energy state possible for any given atom or molecule. However, when a photon strikes a molecule that

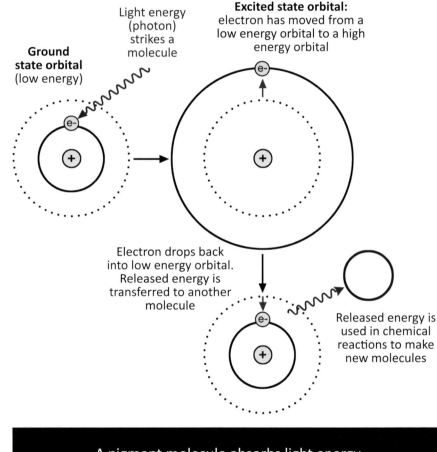

A pigment molecule absorbs light energy

has lots of double bonds and rings, there are high energy states that an electron can occupy. In other words, the molecule can absorb the energy of the photon and use it to move an electron from a low energy state to a high energy state. This electron can stay in the high energy state only briefly, and as it drops back down to the ground state, the energy released can be used to drive chemical reactions.

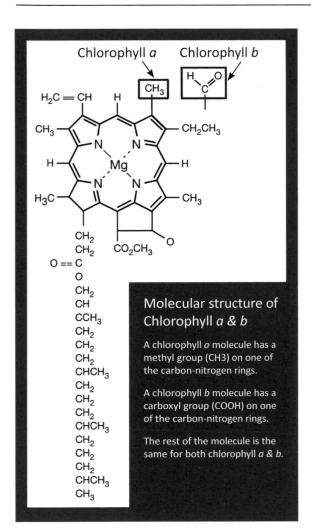

Chlorophyll *a*   Chlorophyll *b*

Molecular structure of Chlorophyll *a* & *b*

A chlorophyll *a* molecule has a methyl group (CH3) on one of the carbon-nitrogen rings.

A chlorophyll *b* molecule has a carboxyl group (COOH) on one of the carbon-nitrogen rings.

The rest of the molecule is the same for both chlorophyll *a* & *b*.

The key light absorbing pigment in most plants is called chlorophyll *a*. Chlorophyll *a* has a connected set of carbon-nitrogen rings attached to a long hydrocarbon tail. An accessory pigment called chlorophyll *b* also absorbs light but at slightly different wavelengths. The only difference in structure between these two molecules is a side-group on one of the carbon-nitrogen rings. Chlorophyll *a* has a methyl group in this position and chlorophyll *b* has a carboxyl group.

Chlorophyll *a* absorbs light in the purple to blue range and in the yellow to red range of the visible spectrum, and chlorophyll *b* absorbs at slightly longer wavelengths in the blue region of the visible spectrum and slightly shorter wavelengths in the red to yellow region. As a result the wavelengths that are reflected make chlorophyll *a* appear blue-green and chlorophyll *b* appear yellow-green in the visible light spectrum.

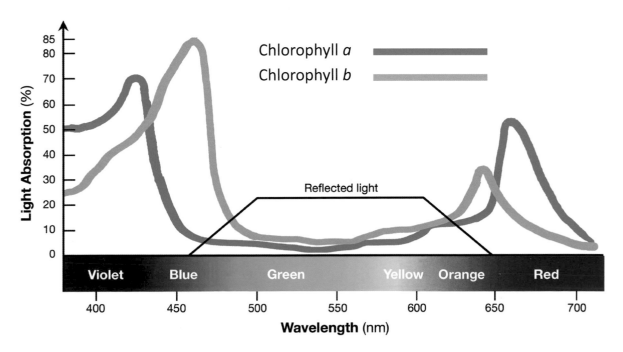

10.3 The Photosystem

How chlorophyll molecules capture and release energy and drive the chemical reactions needed to make sugar involves a complex group of proteins and other molecules working together in a photosystem. There are two photosystems that work together during photosynthesis—photosystem I and photosystem II. Photosystem II functions first in the light-dependent reactions for photosynthesis and is followed by photosystem I.

When light strikes a plant cell, the light is absorbed by a chlorophyll molecule, and this event begins the first series of chemical reactions in photosystem II. Energy from the light is absorbed by the electrons of a chlorophyll molecule and then transferred to other electrons on other chlorophyll molecules. The electrons bounce from molecule to molecule until they reach a specific section of the photosystem called the reaction center complex. Inside the reaction center complex is a special molecule called the primary electron acceptor that takes the electrons, and using the energy released from breaking apart a water molecule, excites them once more to a high energy state. The excited electrons then travel through an elaborate set of proteins called the electron transport chain where they are transferred to photosystem I, and ATP is made. ATP stands for adenosine triphosphate which is a molecule that is essential for providing energy to many chemical reactions inside cells.

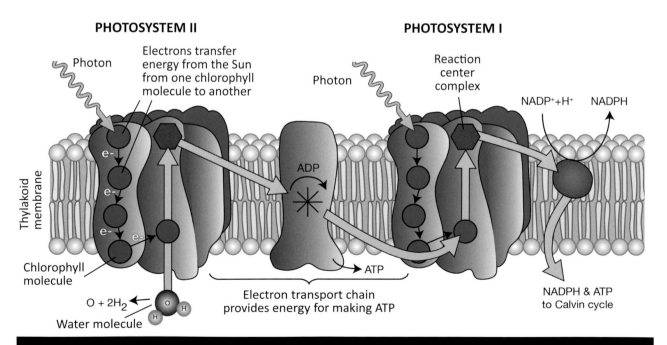

Photosystem II and Photosystem I

A set of chemical reactions similar to photosystem II begins in photosystem I with electrons being absorbed by chlorophyll molecules and eventually transferred to another primary electron acceptor. At this point, when the high energy electrons are transferred to the primary electron acceptor in photosystem I, NADPH is made which is transferred to the Calvin cycle for making sugar. NADPH stands for nicotinamide adenine dinucleotide phosphate, and like ATP, it is also a molecule that provides energy for chemical reactions inside cells.

10.4 The Calvin Cycle

The basic job of photosystem II and photosystem I is to convert the Sun's energy, which arrives in the form of photons, into high energy electrons. The reason the plant cell needs high energy electrons is to make the two very important molecules called ATP and NADPH mentioned above. Both ATP and NADPH are needed in the Calvin Cycle where carbon dioxide is made into sugar.

The Calvin Cycle is a cycle of chemical reactions that can be divided into three separate phases. In phase 1, carbon dioxide enters the cycle and gets converted into 3-phosphoglycerate. In phase 2, ATP enters the cycle, changing 3-phosphoglycerate into

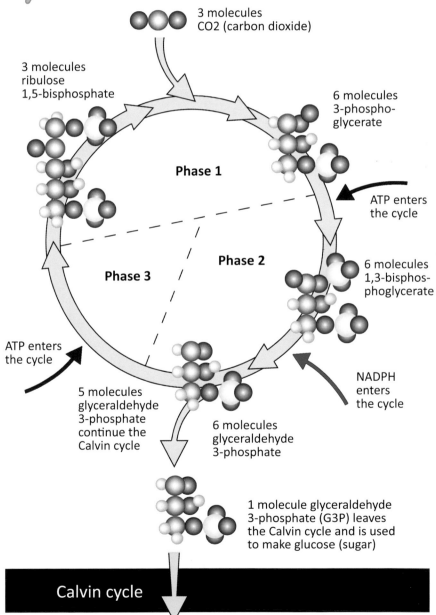

3 molecules
CO_2 (carbon dioxide)

3 molecules
ribulose
1,5-bisphosphate

6 molecules
3-phospho-
glycerate

Phase 1

ATP enters
the cycle

Phase 3

Phase 2

6 molecules
1,3-bisphos-
phoglycerate

ATP enters
the cycle

NADPH
enters
the cycle

5 molecules
glyceraldehyde
3-phosphate
continue the
Calvin cycle

6 molecules
glyceraldehyde
3-phosphate

1 molecule glyceraldehyde
3-phosphate (G3P) leaves
the Calvin cycle and is used
to make glucose (sugar)

Calvin cycle

1,3-bisphosphoglycerate which then gets converted to glyceraldehyde 3-phosphate by NADPH. One of the glyceraldehyde 3-phosphate molecules exits the cycle to make glucose, and the remaining glyceraldehyde 3-phosphate molecules continue through the cycle. In phase 3, ATP enters the cycle to convert glyceraldehyde 3-phosphate into ribulose 1,5-bisphosphate that is used to convert more carbon dioxide into more 3-phosphoglycerate molecules and the process repeats.

In summary, the light-dependent reactions consist of the two photosystems whose job is to convert sunlight into high energy electrons that can be used to make ATP and NADPH. These two molecules are used in the Calvin Cycle which converts carbon dioxide into sugar used by the plant for food.

The Light-dependent Reactions: Photosystems II & I

The Calvin Cycle

The light-dependent reactions and the Calvin cycle in a chloroplast

10.5 Summary

● Plants make their own food by a process called photosynthesis during which light energy, water, and carbon dioxide are used to produce glucose (a simple sugar) and oxygen.

● The process of photosynthesis occurs in plastids called chloroplasts.

● Chlorophyll is the pigment molecule that gives plants their green color by reflecting green light wavelengths.

● Photosynthesis is a series of two distinct stages of multiple chemical reactions known as the light-dependent reactions and the Calvin Cycle. The light-dependent reactions are the chemical reactions that occur when a chloroplast captures light energy from the Sun, and the Calvin Cycle refers to the chemical reactions that produce sugar.

● How chlorophyll molecules capture and release energy and drive the chemical reactions needed to make sugar involves a complex series of proteins and other molecules working together in a photosystem.

● In the Calvin Cycle carbon dioxide is converted to sugar that is used by the plant for food.

10.6 Some Things to Think About

● Why do some plants have broad, flat leaves and some have needles?

● Briefly describe the structure of a chloroplast.

● Explain why plants look green.

● Do you think frogs that are green can perform photosynthesis and make their own food? Why or why not?

● If you wanted to briefly explain how it is that a plant can start from a tiny seed and add enough tissue to grow into a big tree, how would you say it?

● Why are ATP and NADPH important for cells?

● How would you briefly describe what happens during the Calvin cycle?

Chapter 11 Plant Structure and Growth

11.1 Introduction

In the same way that your body has tissues, organs, and cells organized in a particular way that allows you to move, breathe, and walk, plants have tissues, organs, and cells that allow them to grow, to reproduce, and to use the Sun's energy for photosynthesis.

In Chapters 9 and 10 we looked at a generic plant cell and the organelles and plastids that plants utilize within their cells. In this chapter we will take a closer look at some specific types of plant cells and how they are used to make different types of tissues and organs.

Vascular plants have three basic types of organs—roots, stems, and leaves. Each of these organs performs specific functions.

In general, roots have several functions, including absorbing and transporting water

Lilium croceum (botanical illustration)
From the book Addisonia: colored illustrations and popular descriptions of plants, New York Botanical Garden, 1916

and nutrients from the soil, anchoring the plant body to the ground, and helping the plant stay upright. Roots can also serve as a food storage area for the plant. The stem is the part of the plant that supports the plant body, transports nutrients and water, and provides support for the plant to grow tall in search of sunlight. Leaves are the main site of photosynthesis.

11.2 Specialized Cells

In order to perform different functions, organs such as roots, stems, and leaves are made of different tissues with specialized cells. A specialized cell is simply a cell that grows in a specific way to perform a specific function. The process of producing specialized cells is called cell differentiation. Cell differentiation in plants involves changes in the cell wall, the organelles, and the cytoplasm (the water and other molecules that are within the cell but outside the nucleus of the cell).

There are several types of specialized cells—cells that form vascular tissues, cells that form the outer covering of leaves and stems, and cells that provide structural support. Some types of plant cells die at maturity. For example, xylem cells die, forming a hollow interior that allows for the transport of water and minerals through the plant. At the same time, their cell walls provide support for the plant. Other functions within the plant require living cells.

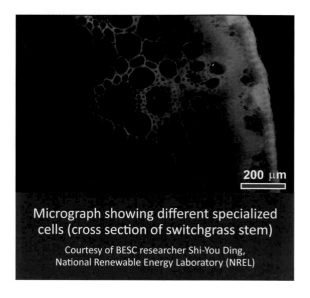

200 μm

Micrograph showing different specialized cells (cross section of switchgrass stem)

Courtesy of BESC researcher Shi-You Ding, National Renewable Energy Laboratory (NREL)

Stems, leaves, and roots have an outer layer of cells called the epidermis. The term epidermis comes from the Greek words *epi* meaning "on" or "above" and *derma* meaning "skin." Epidermal cells form a tightly packed layer of cells on the outside of a plant, and they protect the cells beneath them. Epidermal cells are transparent, allowing light to pass through to the chloroplasts for photosynthesis.

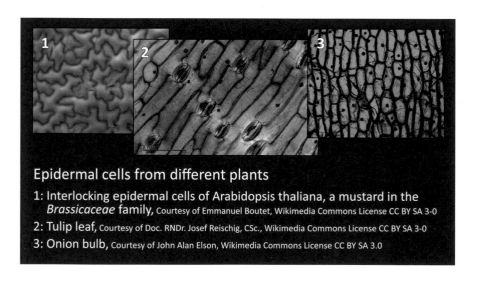

Epidermal cells from different plants

1: Interlocking epidermal cells of Arabidopsis thaliana, a mustard in the *Brassicaceae* family, Courtesy of Emmanuel Boutet, Wikimedia Commons License CC BY SA 3-0
2: Tulip leaf, Courtesy of Doc. RNDr. Josef Reischig, CSc., Wikimedia Commons License CC BY SA 3-0
3: Onion bulb, Courtesy of John Alan Elson, Wikimedia Commons License CC BY SA 3.0

In addition to epidermal cells there are three other main types of cells that make up plant tissues: parenchyma cells, collenchyma cells, and sclerenchyma cells.

Parenchyma cells are the most numerous cells in a plant. They do most of the work of synthesizing and storing molecules. (Synthesis is the process of making molecules through chemical reactions.) In leaves, photosynthesis occurs in parenchyma cells that contain chloroplasts; in stems and roots, parenchyma cells store starch.

Collenchyma cells are elongated with thicker cell walls than parenchyma cells, which helps them provide structural support for plants. The stems of young plants have collenchyma cells just below the epidermis to provide support as the plant grows, and leaf veins have

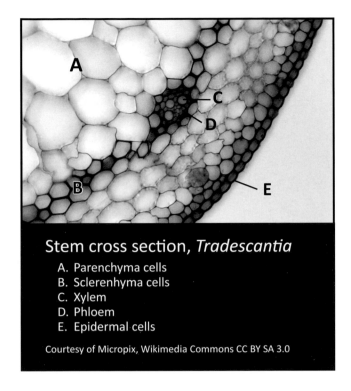

Stem cross section, *Tradescantia*
A. Parenchyma cells
B. Sclerenhyma cells
C. Xylem
D. Phloem
E. Epidermal cells

Courtesy of Micropix, Wikimedia Commons CC BY SA 3.0

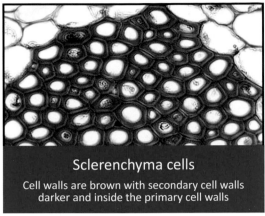

Sclerenchyma cells
Cell walls are brown with secondary cell walls darker and inside the primary cell walls

collenchyma cells along them for strength. Since collenchyma tissues are flexible, they are perfect for giving structure to leaves and young stems. Collenchyma cells stay alive at maturity.

Sclerenchyma cells also support plant structure but are much more rigid than collenchyma cells. Inside their primary cell wall, sclerenchyma cells produce a secondary cell wall that is largely made of cellulose and lignin. Lignin adds strength and rigidity to cell walls and is a major component of wood. Sclerenchyma cells die at maturity, and their primary function is to support the plant with their thick, stiff cell walls.

11.3 Specialized Tissues

In the last section we looked at a variety of specialized cells that make up plant tissues and organs. Now we will take a closer look at specialized tissues. Specialized tissues form from one or more of the specialized cells found in plants. These specialized tissues in turn form the primary organs—roots, stems, and leaves. There are four main types of tissues that make up the primary organs. These tissues are dermal tissue, vascular tissue, ground tissue, and meristematic tissue.

Epidermal cell layer

Epidermal cell layer of *Rumex* leaf
Courtesy of Micropix, Wikimedia Commons CC BY SA 3.0

The dermal tissue system forms on the outside of the plant and acts like the "skin" covering the plant body. In non-woody plants, dermal tissue is composed of a single layer of epidermal cells that protect the plant from physical damage and insects. In woody plants a protective tissue called the periderm replaces the epidermis in older regions of the stem and roots. The periderm has more than one layer of cells that are dead at maturity and form bark on trees.

The vascular tissues, xylem and phloem, make up the vascular tissue system of the plant body. The vascular tissue system acts like the "bloodstream" of the plant, moving nutrients and water up through the xylem from the roots to the stems and leaves, and sugars down through the phloem to the leaves, stems, and roots.

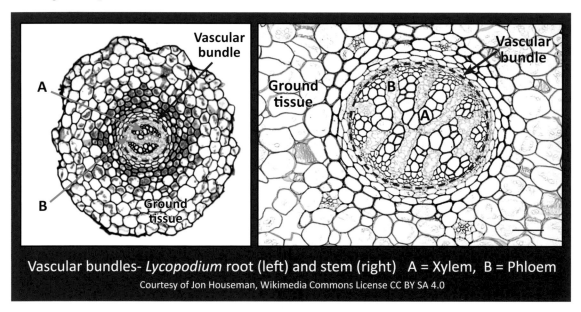

Vascular bundles- *Lycopodium* root (left) and stem (right)   A = Xylem, B = Phloem
Courtesy of Jon Houseman, Wikimedia Commons License CC BY SA 4.0

Ground tissue is found inside the dermal tissue and around the vascular tissues. It makes up the bulk of the plant and has a number of functions including photosynthesis, storage of large molecules, and support. Ground tissue is composed of parenchyma cells, collenchyma cells, and sclerenchyma cells. Pith is ground tissue found in the center section of a stem and is used for food storage.

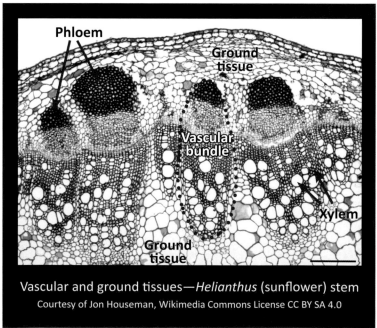

Vascular and ground tissues—*Helianthus* (sunflower) stem
Courtesy of Jon Houseman, Wikimedia Commons License CC BY SA 4.0

The meristem is tissue made of undifferentiated cells called meristematic cells. Meristematic cells are the only cells in a plant that can divide and allow the plant to grow. All the new cells in a plant are produced by meristematic cells. Because they are undifferentiated,

meristematic cells can produce any kind of cell that is needed by the plant. All of the cell types we have looked at originate in the meristematic tissue. When the meristematic cells begin to divide, the new cells look alike, but they differentiate into one of the three main cell types described in Section 11.2.

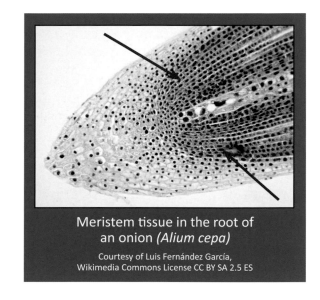

Meristem tissue in the root of an onion *(Alium cepa)*

Courtesy of Luis Fernández García, Wikimedia Commons License CC BY SA 2.5 ES

11.4 Organs: Roots

Roots are the underground organ of a plant. Roots anchor plants to the soil and absorb nutrients and water. When a seed has the right conditions available, it starts to grow, sending roots into the surrounding soil. From these first roots the plant continues to grow into a seedling and eventually a mature plant.

Fibrous roots of the monocot *Chlorophytum comosum* (spider plant)

Courtesy of Wildfeuer, Wikimedia Commons CC BY SA 3.0

We saw in Chapter 9 that there are two main types of root systems for plants: fibrous roots and taproots. Fibrous roots are found primarily in monocots, and taproots are found primarily in dicots. Some taproots, such as beets and carrots, are major food storage areas for the plant and are edible.

Red beets are edible taproots

Courtesy of Evan Amos

How do roots absorb nutrients from the soil? A root cannot just sit passively in the soil and soak up water. Instead, it must actively transport water and nutrients into and through its tissues into the xylem. This process begins with osmosis, and roots create the conditions necessary for osmosis to occur. Osmosis happens when molecules move from an area of high concentration to an area of low concentration through a permeable or semipermeable membrane. Because epidermal cells in roots are surrounded by a semipermeable membrane, roots can uptake water and mineral ions from the soil. When the concentration of water molecules or mineral ions outside the root is higher than inside the root, water or ions from the soil will flow into the epidermal tissues in an effort to equalize the concentration of water and minerals inside and outside the root. As these molecules are transferred through the root to the xylem, the concentration in the root once again becomes lower inside the root than outside, and more molecules will enter through the membrane. In this way roots effectively pump water into their tissues.

The water and minerals that have entered the root's epidermal tissues then move toward the central portion of the root and into the ground and vascular tissues. Water and mineral ions can travel either between the cells through microscopic channels in the cell walls or through the inside of the cells, going through cell membranes from one cell to the next. Molecules taking the route through the cells move across cell walls by means of microscopic channels in the cell walls. These channels are called plasmodesmata (singular, plasmodesma).

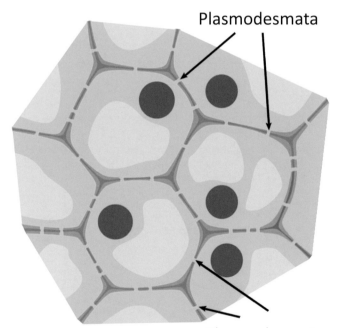
Plasmodesmata

Plasmodesmata

11.5 Organs: Stems

Stems have three important functions. They transport nutrients between the roots and leaves; produce leaves, branches, and flowers; and hold the leaves upright and at angles that optimize exposure to sunlight. Stems are made of dermal, vascular, and ground tissues. Recall from Section 11.3 that dermal tissues make up the external layer, vascular tissues

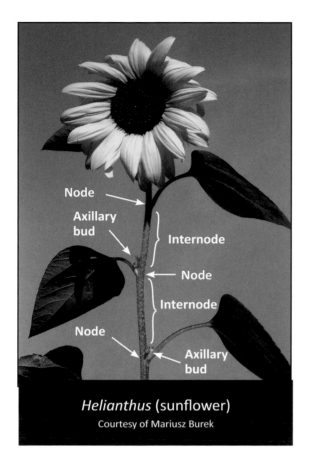

Helianthus (sunflower)
Courtesy of Mariusz Burek

transport nutrients, and ground tissues are used for photosynthesis and storage of large molecules.

If you look closely at a plant stem, you can see distinct places from which leaves grow. These places along the stem are called nodes. At the node, the angle between the stem and leaf is called the axil, and the axil is where axillary buds (lateral buds) are formed. Axillary buds can grow to become stems or flowers.

The part of the stem in between two nodes is called the internode. The length of the internodes is partly responsible for whether the plant is short or tall.

As we saw in Chapter 9, there are two main types of vascular plants—monocots and dicots. Monocot and dicot stems differ in their organization of both vascular and ground tissues. The vascular tissues of monocot stems are bundled together with the xylem toward the center of the stem and the phloem toward the outside.

These vascular bundles are scattered throughout the ground tissue of the monocot stem. The ground tissue is made primarily of parenchyma cells.

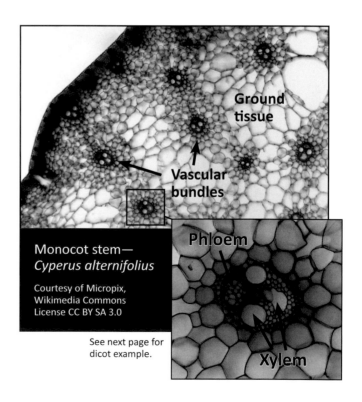

Monocot stem—
Cyperus alternifolius

Courtesy of Micropix,
Wikimedia Commons
License CC BY SA 3.0

See next page for dicot example.

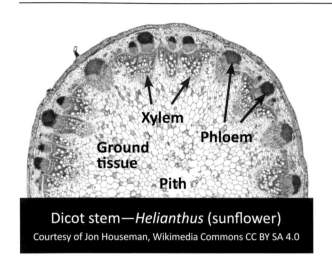

Dicot stem—*Helianthus* (sunflower)
Courtesy of Jon Houseman, Wikimedia Commons CC BY SA 4.0

The vascular tissues of dicot stems are localized near the epidermal tissues and arranged in a ring-like pattern. The pith makes up the central portion of a dicot stem and the ground tissue is made primarily of parenchyma cells.

Stems increase in length by adding cells to the ends of mature parts of the stems. This new growth is called a shoot. (The entire above-ground part of the plant is also called the shoot or the shoot system.) This type of growth is called primary growth. Seeded plants undergo primary growth throughout their entire life. Primary growth occurs in the apical meristem tissue located at the very tips of shoots and roots. In stems the apical meristem is a dome-shaped structure where cells begin to divide and multiply.

As the stem continues to increase in length, it will also need to increase in diameter in order to support continued growth. Plant stems increase in diameter through a process called secondary growth. Secondary growth is produced by lateral meristems located along the length of stems. The lateral meristems are composed of two areas called the vascular cambium and the cork cambium. The vascular cambium adds new xylem and phloem tissues, and the cork cambium adds new dermal tissue to the stem's exterior. Secondary growth also occurs in roots.

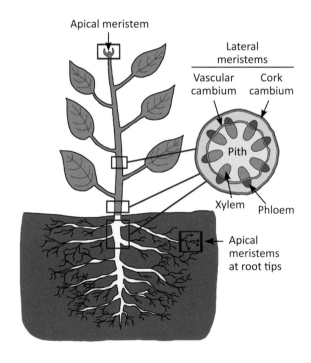

11.6 Organs: Leaves

The primary organ that performs photosynthesis is the leaf. Although leaves come in many different shapes and sizes, they all are optimized for the absorption of light and the exchange of gases with the atmosphere. Leaves also release heat and defend the plant from predators and pathogens (disease causing microorganisms). To best meet these demands in different environments, leaf shapes vary as does the placement of leaves on the stem.

The broad, flattened part of the leaf is called the blade. In many plants the blades are attached to the plant with a petiole—a stalk that attaches the blade to the stem at a node. Other plants, for example corn and other grasses, have leaves that are attached directly to the stem. Veins contain the vascular tissues of a leaf and also provide support for the leaf, helping it to maintain its shape.

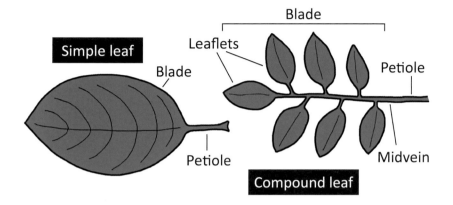

A simple leaf has one undivided blade. A compound leaf is divided into several leaflets.

The way leaves grow from nodes helps with plant identification. Some plants have one leaf per node, which is called an alternate arrangement, some have two leaves per node, which is an opposite arrangement, and others have three or more per node, which is a whorled arrangement.

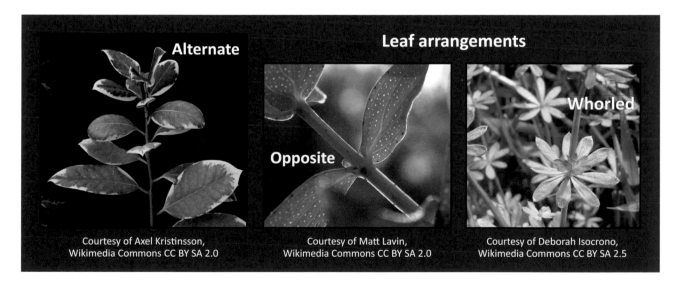

Leaf structure can help identify the plant as a dicot or a monocot. A dicot leaf generally has a blade that is attached to a petiole with the veins arranged in a net-like pattern. Monocots such as grasses do not have a petiole; instead, the base of the blade wraps around the stem. The veins in a monocot leaf have a parallel arrangement.

11.7 Summary

● Vascular plants have three organ types: roots, stems, and leaves.

● Different types of specialized cells are required to make up the various organs in vascular plants.

● Specialized tissues form from one or more of the specialized cells found in plants. These specialized tissues in turn form the primary organs—roots, stems, and leaves.

● Roots anchor plants to the ground and absorb from the soil the water and nutrients needed for use in photosynthesis.

● Stems transport nutrients between the roots and leaves; produce leaves, branches, and flowers; and hold the leaves upright and at angles that optimize exposure to sunlight.

● Leaves are the primary organs that perform photosynthesis.

11.8 Some Things to Think About

● Why do you think plants need to have different types of cells?

● Briefly describe epidermal cells, parenchyma cells, collenchyma cells, and sclerenchyma cells and the functions that each type of cell performs.

● What is the difference between xylem and phloem?

● What are the four main types of tissues in a plant and what does each one do?

● Briefly describe the two pathways by which water travels through a root.

● What are the primary functions of a plant stem?

● How do stems grow longer and increase in diameter?

● What features of leaves could you use to help identify a plant?

Chapter 12 Plant Reproduction

Portulaca oleracea (common purslane)
Courtesy of Didier Descouens, Wikimedia
Commons License CC BY SA 4.0

12.1 Introduction

Reproduction is a vital activity for all organisms, including plants. If plants cannot successfully pass down their genetic information to a new generation, they will become extinct. And just like there are endangered animals, there are also endangered plants that may soon cease to exist. For example, the yellow coneflower, which was once an abundant native species west of the Mississippi River, has dwindled in numbers over the last several decades. One of the most endangered plant species on the planet is the Texas wild rice with less than 150 clumps remaining. This plant grows in the fresh water of the San Marcos River, and its numbers are dwindling due to lowered water levels caused by a nearby dam and human water use. Among many other endangered plant species are the Arizona agave and the Georgia aster.

As we saw in Chapter 9, plants use a variety of methods to reproduce and pass on their genetic information. In this chapter we will take a closer look at these methods and explore in detail asexual and sexual reproduction methods of non-seeded plants, non-flowering seeded plants, and flowering seeded plants.

12.2 Asexual Reproduction

Many plants can reproduce asexually, producing new plants that are clones of the parent plant. Clones are genetically the same as the parent. Recall from Chapter 9 that asexual reproduction in plants is called

A few plant species that are endangered in the wild

Photo credits: 1. *Escobaria minima*, Andrea Miclet, CC BY SA 2.0; 2. Georgia aster, *Symphyotrichum georgianum*, Biosthmors, CC BY SA 4.0; 3. Yellow coneflower, *Echinacea paradoxa*, Phyzome, CC BY SA 3.0; 4. *Agave arizonica*, Stan Shebs, CC BY SA 3.0; 5. *Gardenia brighamii*, Forest & Kim Starr, CC BY SA 3.0

vegetative propagation. In this section we'll take a look at some of the common methods of vegetative propagation in both nonvascular and vascular plants.

Nonvascular Plants

In Chapter 9 we learned that nonvascular plants include all the mosses, hornworts, and liverworts.

Phylum

Nonvascular Non-seeded
- Bryophyta - mosses
- Anthocerotophyta - hornworts
- Marchantiophyta - liverworts

One type of vegetative propagation used by nonvascular plants to reproduce is fragmentation, which occurs when part of a plant breaks off to grow a new plant. In nonvascular plants, pieces of the plant can simply break off and grow into new plants. These pieces can be small bits of leaves or whole sections of the plant. When the plant part breaks off and falls in a suitable location that has plenty of water, it will begin to form a new plant. To aid the fragmentation process, some plants create areas in their leaves or stems that are very fragile and easy to break. This allows sections of the plant to be broken off by a breeze, insect, or small animal, giving the plant a way to grow a new individual in a new location.

A piece of a moss breaks off and falls to the ground

The plant fragment begins to grow into a new plant

Vascular Non-seeded Plants

Recall that the vascular non-seeded plants include the club mosses, horsetails, ferns, and whisk ferns. Vascular non-seeded plants can reproduce by some types of vegetative propagation, forming clones of themselves. For example, many vascular non-seeded plants produce new plants from their

Phylum

Vascular Non-seeded
- Lycopodiophyta - club mosses
- Sphenophyta - horsetails
- Polypodiophyta - ferns
- Psilophyta - whisk ferns

rhizomes. Some fern fronds sprout plantlets, or baby plants, from the margins (edges) of their fronds. When plantlets or entire fronds drop to the soil, the plantlets can take root and form new plants. Some club mosses form small bulb-shaped buds called bulbils for asexual reproduction. A bulbil is produced in a leaf axil, and when it has developed enough to grow into a new plant, the bulbil falls to the ground and roots in the soil.

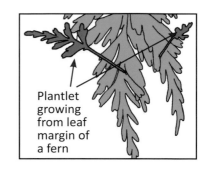

Plantlet growing from leaf margin of a fern

Vascular Seeded Plants

Vascular seeded plants can also reproduce through vegetative propagation using stems, leaves, or roots.

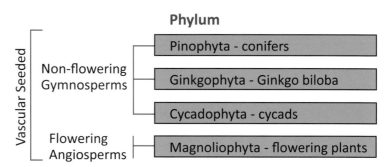

Phylum

Vascular Seeded

Non-flowering Gymnosperms
- Pinophyta - conifers
- Ginkgophyta - Ginkgo biloba
- Cycadophyta - cycads

Flowering Angiosperms
- Magnoliophyta - flowering plants

Vegetative Propagation Using Stems

Stems are the part of the plant that is above ground, providing support for the plant and its leaves, but many types of vegetative propagation involve modified stems that grow under or along the surface of the ground. If you dig up something that looks like a root but has nodes (see the figure below and refer to Chapter 11), you can tell that it's an underground stem because roots don't have nodes. A new clone can grow from a node on an underground stem. Modified stems include rhizomes, tubers, and stolons.

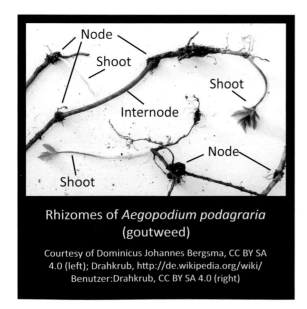

Node
Shoot
Shoot
Internode
Node
Shoot

**Rhizomes of *Aegopodium podagraria*
(goutweed)**

Recall from Chapter 9 that rhizomes are underground, horizontal stems used by plants for food storage. Rhizomes have nodes and short internodes and often have small, scalelike leaf structures.

A tuber is a large thickened part of a rhizome where a plant can store a lot of nutrients to use for growing new plants. One plant can produce many tubers in a growing season. The potato is a typical tuber and grows at the tip of a rhizome. By looking closely at a potato, you can see the features of a tuber. Bud structures called eyes are scattered over the surface. Each eye has a half-moon shaped ridge that is the remains of a scale leaf, which is a small modified leaf structure that is often thin and transparent. Below this ridge is a bud that can grow into a new plant. The interior of the potato contains carbohydrates to be used for food by a newly sprouting plant.

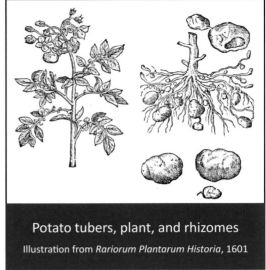

Potato tubers, plant, and rhizomes
Illustration from *Rariorum Plantarum Historia*, 1601

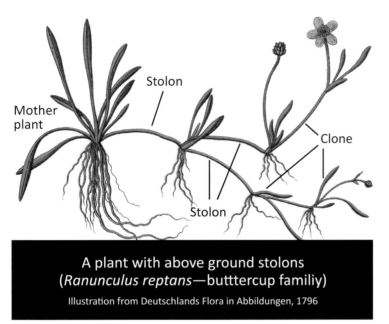

A plant with above ground stolons
(*Ranunculus reptans*—butttercup familiy)
Illustration from Deutschlands Flora in Abbildungen, 1796

Stolons, also called runners, are modified stems that grow above ground or just below the surface and spread out horizontally from the stem of the mother plant. New shoots are produced from nodes, and stolons often have long internodes. Stolons can be found on plants such as the strawberry where you might find a mother plant surrounded by genetically identical daughter plants that have sprouted from nodes on the stolons. As with rhizomes, a node can form a new plant even if it is detached from the rest of the stolon. The distinction between stolons and rhizomes can sometimes be difficult to determine, especially for stolons that grow just below the surface of the soil. The spider plant is a popular houseplant that grows plantlets dangling from stolons.

Vegetative Propagation Using Leaves

Bulbs are structures made of layers of modified leaves. A bulb is usually large and globe-shaped, grows underground, and provides food storage for the plant to help it survive unfavorable conditions, such as winter weather and drought. During these unfavorable times, the plant becomes dormant, or inactive. When conditions for growth improve, the plant can use food stored in the bulb to begin to grow again. Inside the layers of the bulb is a bud that can grow into the above ground shoot system of the plant. An onion is an example

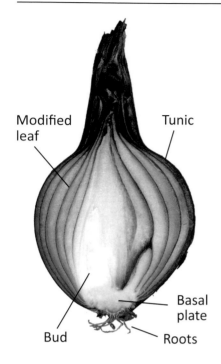

Modified leaf

Tunic

Basal plate

Bud

Roots

of a typical bulb. If you look at an onion, at the bottom you can see a very short, flattened stem structure called the basal plate from which roots grow downward and the modified leaves grow upward. A bulb is covered by a papery tunic, and when cut open, you can see that a bulb is layered inside. Garlic is another example of an edible bulb. Each clove is a separate bulb that is capable of forming a new plant, and if you cut a garlic clove in half, you can see the layered structure.

Like some ferns, some types of vascular seeded plants form plantlets that develop from meristematic tissue on the margin of their leaves. When the plantlet grows to be a size where it can live on its own, it falls off the plant, and if it lands in a good location, it can take root and become a new plant.

Plantlets on leaf margin
Courtesy of CrazyD, Wikimedia Commons License, CC BY SA 3.0

Vegetative Propagation Using Roots

Pando *(Populus tremuloides)*
Courtesy of J. Zapell, USDA

Suckers are new plants that sprout from meristematic tissue found in buds on roots. For example, aspen trees often reproduce vegetatively by growing new trees from one root system, and an entire grove of aspen trees can be genetically identical individuals growing from the same root system. A famous example is the Pando aspen colony in Utah, covering over 40 hectares (100 acres). Pando has over 40,000 tree trunks growing from one root system and is estimated to be 12,000-80,000 years old or maybe even older!

Another example of using roots for propagation is seen in plants that have fleshy roots or tuberous roots. Fleshy roots are those that have thickened taproots, such as carrots and beets. Tuberous roots have thickened areas on secondary roots that branch out laterally from the main root. The sweet potato is an example of a tuberous root. Because they are made of root tissue, fleshy roots and tuberous roots don't have nodes, buds, or scale leaves. Sweet potatoes, dahlias, and jerusalem artichokes can grow new plants from the tuberous

Sweet potato—tuberous roots and plant *(Ipomoea batatas)*
Courtesy of Miya, CC BY SA 2.1 Japan

part of the root. So, even though they both grow underground, a white potato is a true tuber and is part of a modified stem (a rhizome), while a sweet potato is part of the root of a plant.

12.3 Sexual Reproduction

Although different types of plants vary in the details of how they reproduce sexually, there are some basics common to all plants. The sexual reproduction cycle, also called the alternation of generations, includes two stages—the sporophyte stage and the gametophyte stage—with each stage being referred to as a generation. The stages cycle (alternate), with one giving rise to the other.

The terminology gets a bit complicated. Following is a list of some word elements and definitions to help you decipher some of these terms. Rather than memorizing the terms, it's more important to understand the different steps in the sexual reproductive cycles of plants.

The Sporophyte Stage

The sporophyte stage can be thought of as the spore-producing stage of the sexual reproductive cycle. During this stage the plant produces the cells, called spores, that will develop into male and female reproductive cells. In the sporophyte stage, the cells in the plant are diploid meaning they contain the usual two sets of chromosomes that carry the genetic information of the plant. At this stage in the life cycle, the plant itself is called a sporophyte.

The sporophyte uses a process called meiosis to produce cells that have only half the usual number of chromosomes. A parent cell has two sets of chromosomes (a full set),

Astounding Botanical Terminology!

Although the terminology in botany can seem like gobbledygook, scientists try to precisely name each plant part and function. Because plants are such complicated organisms, the terminology also gets complicated, especially as you begin to look more deeply into how plants function. Many scientific terms contain elements that come from Greek or Latin. Knowing what the elements mean can help you decipher the term. For example, knowing that *phyte* means plant and *phyll* means leaf, you can begin to understand the meaning of a term that contains one of these elements.

Word Elements

angi- = vessel	Gr., *angeion*
arche- = beginning	Gr., *archein* (to begin)
di- = two	Gr., *di-*
diplo- = double	Gr., *diploos* (double)
epi- (ep- or eph-) = on, upon, above	Gr., *epi*
gamete (gameto-) = reproductive cell	Gr., *gamein* (to marry)
gon- = reproductive structure	Gr., *gonos* (offspring, seed, that which engenders)
haplo- = single	Gr., *haploos* (single, simple)
-ium (-ia) = small	Gr., *idion* (suffix meaning small)
macro = large	Gr., *makros*
mega = large	Gr., *megas*
micro = small	Gr., *mikros*
-oid = form	Gr., *eidos*
ova- (ovu-) = egg, female reproductive part	L., *ovum*
-phyll = leaf	Gr., *phyllon*
-phyte, phyto- = plant	Gr., *phyton*
sperm (sperma-) = sperm, male reproductive part	L., *sperma* (seed)
spore (sporo-) = reproductive cell of a plant	Gr., *spora* (seed)

Some Terms Containing Word Elements

archegonium (är-ki-gō'-nē-əm) [*pl.,* archegonia] • in certain plants, a small female structure in which egg cells are formed [*archein* (beginning) + *gonos* (seed or reproductive structure) + *-ium* (small)]

diploid (di'-ploid) • having two sets of chromosomes [*diploos* (double) + *eidos* (form)]

gametophyte (gə-mē'-tə-fīt) • gamete-producing stage of a plant [gamete (reproductive cell) + *phyte* (plant)]

haploid (ha'-ploid) • having one set of chromosomes [*haploos* (double) + *eidos* (form)]

megasporangium (me-gə-spə-ran'-jē-əm) [*pl.,* megasporangia] • structure that produces female spores [*mega-* (large) + *spora* (seed) + angi- (vessel) + *-ium* (small)]

megaspore (me'-gə-spôr) • a single cell (spore) that will develop into the female gamete (egg); is usually larger than the microspore [*mega-* (large) + *spora* (seed)]

microsporangium (mī-krə-spə-ran'-jē-əm) [*pl.,* microsporangia] • structure that produces male spores [*micro-* (small) + *spora* (seed) + *angi-* (vessel)+ *-ium* (small)]

microspore (mī'-krə-spôr) • a single cell (spore) that will develop into the male gamete (sperm); is usually smaller than the megaspore [*micro-* (small) + *spora* (seed)]

sporangium (spə-ran'-jē-əm) [*pl.,* sporangia] • a structure in which spores are produced [*spora* (seed) + *angi-* (vessel)]

sporophyte (spôr'-ə-fīt) • the spore-producing stage of a plant [*sporo-* (seed) + phyte (plant)]

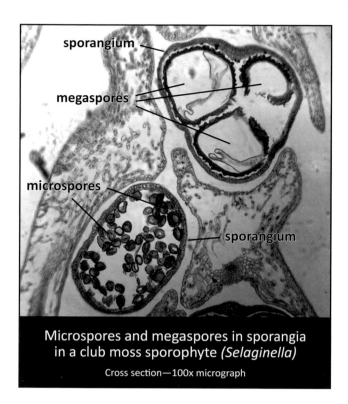

sporangium

megaspores

microspores

sporangium

Microspores and megaspores in sporangia
in a club moss sporophyte *(Selaginella)*

Cross section—100x micrograph

and each daughter cell gets one set of the chromosomes (½ of the full set). Because daughter cells have only one set of chromosomes, these cells are called haploid.

The structure in which the haploid spores are produced is called a sporangium. There are two types of haploid cells produced in sporangia. The megaspore is a single cell that will develop into a female egg cell and is produced in a megasporangium. The microspore is a single cell that will develop into a male sperm cell and is produced in a microsporangium. Egg cells and sperm

cells are called gametes. Note that the prefix mega- means "large" and micro- means "small." The female megaspores are generally larger than the male microspores. These spore cells are not capable of individually forming new plants. Also note that some plant species have separate male and female plants, with one plant producing either male or female spores but not both. Other plants produce both male and female spores on the same plant.

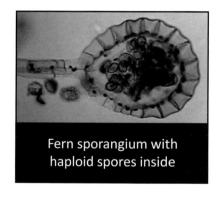

Fern sporangium with
haploid spores inside

The Gametophyte Stage

Haploid cells produced by the sporophyte grow to make the structure of the gametophyte. During the gametophyte stage, gametes are produced from the spore cells, and cell division occurs by mitosis. Recall that mitosis is a type of cell division in which the new cells have the same number of chromosomes as the cell they were formed from—the parent cell. With mitosis, haploid cells will divide to produce more haploid cells, and diploid cells will divide to produce more diploid cells. Therefore, cells produced by the gametophyte will also be haploid like their parent cells. Megaspores will develop into haploid female gametes (eggs), and microspores will develop into haploid male gametes (sperm). When an egg cell and a sperm cell unite, they form a zygote, a cell that contains genetic material from both the sperm cell and the egg cell. The zygote is diploid, containing two sets of chromosomes, and when it divides, the new cells will also be diploid. The union of the sperm and egg is called fertilization. The zygote will eventually turn into a seed or a structure that is able to grow into a new plant that is genetically different from its parents. A new sporophyte stage begins with the production of the diploid cells.

Nonvascular Plants

Nonvascular plants reproduce sexually by creating haploid spores, which in this usage of the word are single cells that are capable of growing into a new plant. In general, if soil and weather conditions are favorable, mosses,

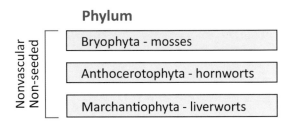

Phylum

Nonvascular Non-seeded

Bryophyta - mosses

Anthocerotophyta - hornworts

Marchantiophyta - liverworts

liverworts, and hornworts will use sexual reproduction to propagate. These plants reproduce sexually in similar ways, but the size and shape of their reproductive structures differ.

Liverwort *(Marchantia polymorpha)*
Left: Top view of female gametophytes; Right: Male gametophytes
Courtesy of Plantsurfer-CC BY SA 3.0

In nonvascular non-seeded plants the gametophyte stage is dominant—it is the plant you see and identify as a moss, a liverwort, or a hornwort. At certain times you may see the spore-containing capsules of a sporophyte that is growing on the gametophyte.

Vascular Non-seeded Plants

Vascular non-seeded plants also reproduce sexually, creating spores that can grow into new plants. Similar to nonvascular non-seeded plants, in the gametophyte stage the spores are produced when a sperm cell combines with an egg cell, forming a zygote. The zygote develops into the sporophyte stage of the plant.

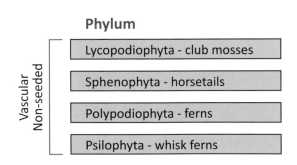

Phylum

Vascular Non-seeded

Lycopodiophyta - club mosses

Sphenophyta - horsetails

Polypodiophyta - ferns

Psilophyta - whisk ferns

Let's look at ferns for an example of sexual reproduction in vascular non-seeded plants. The fern fronds you find growing in shady places are the sporophyte, or spore-bearing, diploid stage of the plant. As in other plants, spores are formed in structures called sporangia. In a fern, the sporangia are found on the underside of the fronds and are clustered in clumps called sori (singular, sorus).

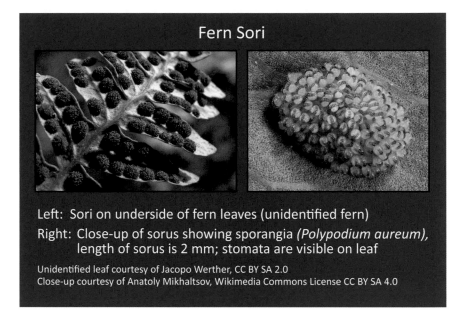

Fern Sori

Left: Sori on underside of fern leaves (unidentified fern)
Right: Close-up of sorus showing sporangia *(Polypodium aureum),* length of sorus is 2 mm; stomata are visible on leaf

Unidentified leaf courtesy of Jacopo Werther, CC BY SA 2.0
Close-up courtesy of Anatoly Mikhaltsov, Wikimedia Commons License CC BY SA 4.0

When the spores in the sori mature, they're released into the air, in some species explosively as a sorus bursts. If a spore lands in soil and the conditions are right, it will grow into a tiny, often heart-shaped, plantlet that will grow into a mature fern.

Vascular Seeded Plants

Recall from Chapter 9 that vascular seeded plants include both non-flowering gymnosperms and flowering angiosperms. Vascular seeded plants are the largest group of plants because they are adaptable to many different environments.

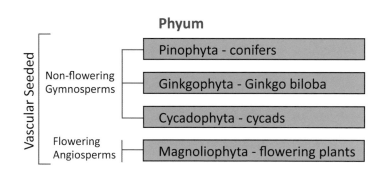

Phyum

Vascular Seeded

Non-flowering Gymnosperms
- Pinophyta - conifers
- Ginkgophyta - Ginkgo biloba
- Cycadophyta - cycads

Flowering Angiosperms
- Magnoliophyta - flowering plants

Why Seeds?

So far in this chapter, we have explored how both nonvascular non-seeded and vascular non-seeded plants reproduce asexually using vegetative propagation and sexually using spores. One of the major disadvantages non-seeded plants face in reproducing is their need for plenty of water. This limits the geographical habitats of non-seeded plants to those environments with a lot of moisture, such as rivers, ponds, marshes, and damp woods. But what about plants growing on mountaintops or in the middle of desert environments? How do these plants thrive in the absence of abundant water?

The answer is seeds! A seed is a plant embryo enclosed in a protective coating. Seeds allow a plant to reproduce without an abundance of water. The seed coat protects the vital genetic material in a seed, allowing a plant to keep its genetic material dormant until the conditions are just right for germination. Also, a seed provides food for the plant embryo, allowing it to establish growth before requiring nutrients from the surrounding environment.

There are two types of seeded plants: those that produce flowers and those that do not produce flowers. We will take a closer look at each of these in the following sections.

Gymnosperms

We saw in Chapter 9 that seeded non-flowering plants are called gymnosperms. The word gymnosperm comes from two Greek words—*gumnos* meaning "naked" and *sperma* meaning "seed." Gymnosperms are thought of as plants having "naked seeds" because the seeds of gymnosperms are not surrounded by fruit like the seeds of angiosperms are; instead, the seeds are

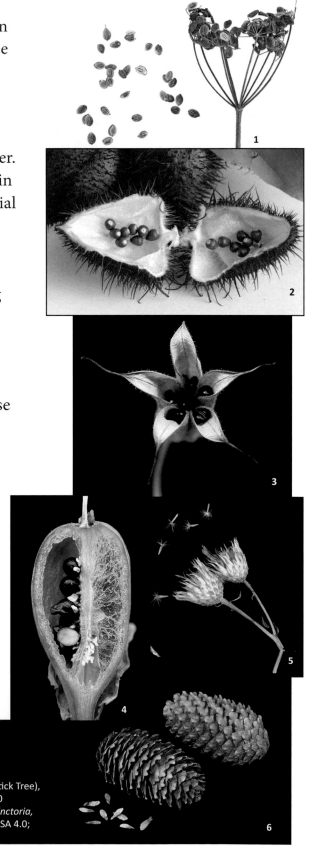

Seeds with their seed-bearing structures

Photos credits (Wikimedia Commons Licenses)

1. *Pastinaca sativa*, Didier Descouens, CC BY SA 4.0; 2. *Bixa orellana* (Lipstick Tree), Challiyan, CC BY SA 3.0; 3. *Aquilegia vulgaris*, Frank Vincentz ,CC BY SA 3.0
4. *Narcissus pseudonarcissus*, Frank Vincentz, CC BY SA 3.0; 5. *Serratula tinctoria*, Roger Culos, CC BY SA 3.0; 6. *Picea likiangensis*, Didier Descouens, CC BY SA 4.0;

exposed. The sexual reproductive structures of gymnosperms are cones rather than flowers. Most gymnosperms have their seeds arranged on the outside of a female cone. The cone may have scales that cover and protect the seeds while they are maturing, but the seed itself does not have an outer covering of fruit. Although gymnosperms reproduce mainly through sexual reproduction, some also use a few methods of vegetative propagation, such as suckering.

Some gymnosperm seeds have a seed scale attached to catch the wind
Pinus sylvestris (Scots pine)
Courtesy of Beentree, Wikimedia Commons CC BY SA 3.0

In gymnosperms, and also in angiosperms, the sporophyte stage is dominant—it's the main part of the plant that you see. When you look at a forest of conifers, you are seeing the sporophytes. In the sporophyte stage, microspores and megaspores are produced in sporangia located in the cones. Microspores and megaspores develop into gametes in separate male and female cones, beginning the gametophyte stage. The gametophyte stage ends with the formation of seeds. For gymnosperms the gametophytes (cones) grow on and are nurtured by the sporophyte.

Ovulate cone
Pinus longaeva (bristlecone pine)
Courtesy of Dcrjsr, CC BY SA 3.0

Pollen cones—*Pinus taeda*
(Loblolly Pine)
Courtesy of Rror, CC BY SA 3.0

Gymnosperms have two types of cones—a female cone and a male cone. The female cone is called the ovulate cone or seed cone. The male cone is called the pollen cone. Ovulate cones are generally much larger than pollen cones.

On the pollen cones, the microspores develop into pollen grains. A pollen grain is microscopic and generally is made of four cells. Each grain of pollen contains a haploid male gamete (sperm cell). When the pollen grains are released by the pollen cone, wind carries the pollen to the ovulate cones. Conifers release a huge quantity of pollen, and at the right time of year areas near conifers may be covered with yellow dust made of pollen grains.

While the male cones are forming pollen, the ovulate cones are forming ovules. The female gamete (egg) develops in the megasporangium. Pollination occurs when a pollen grain lands on the end of the ovule and the pollen grows a pollen tube that enters the ovule through a tiny hole in the protective covering and grows through the ovule until it reaches the egg. At this time the male and female gametes unite, forming a diploid zygote that will grow into a small embryo that develops into a seed. The seed is made up of the embryo, surrounding tissues that contain stored food, and a protective seed coat. When fully developed, the seed is released by the ovulate cone, and under the proper conditions it will grow into a new sporophyte seedling.

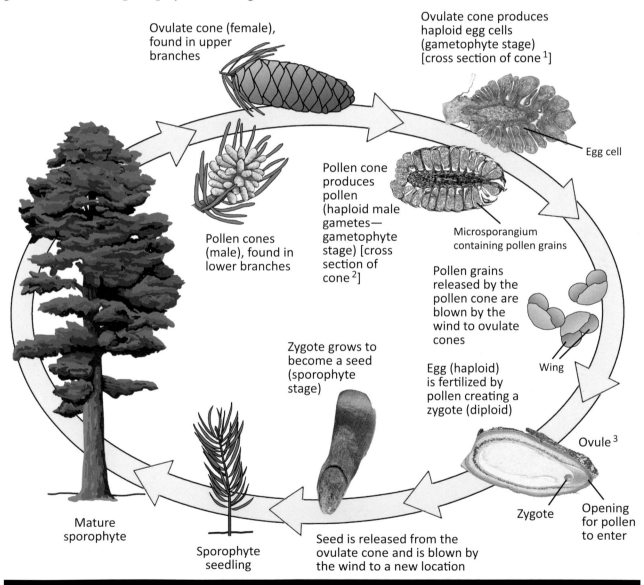

Ovulate cone (female), found in upper branches

Ovulate cone produces haploid egg cells (gametophyte stage) [cross section of cone [1]]

Egg cell

Pollen cone produces pollen (haploid male gametes—gametophyte stage) [cross section of cone [2]]

Pollen cones (male), found in lower branches

Microsporangium containing pollen grains

Pollen grains released by the pollen cone are blown by the wind to ovulate cones

Zygote grows to become a seed (sporophyte stage)

Egg (haploid) is fertilized by pollen creating a zygote (diploid)

Wing

Ovule [3]

Mature sporophyte

Sporophyte seedling

Seed is released from the ovulate cone and is blown by the wind to a new location

Zygote

Opening for pollen to enter

Reproductive Cycle of a Gymnosperm (conifer—*Pinus sp.*)

[1-3] Micrographs courtesy of Jon Houseman, Wikimedia Commons License CC BY SA 4.0

The reproductive cycle can be quite lengthy in gymnosperms and may take two years or more to complete. In some pine tree species (genus *Pinus*) it can take 15 months for the female gametophyte to mature and about the same length of time for the male gamete to reach the egg cell. Additional time is needed for the seed to mature, fall to the ground, and grow a new plant.

Angiosperms

Like gymnosperms, angiosperms reproduce using seeds. However, unlike gymnosperms where movement of both pollen and seeds is accomplished primarily by wind, angiosperms utilize butterflies, bees, birds, and animals of all sizes to spread both pollen and seeds to new locations. In order to get butterflies, bees, and birds interested in doing the work of transporting pollen, the plant offers a "reward" in the form of nectar in flowers. When insects eat the nectar, pollen sticks to their bodies and in this way they carry the pollen from one flower to another, spreading the angiosperm's genetic material to other plants of the same type.

A honeybee dusted with pollen

The yellow structure on the hind leg is a pollen sac that the bee fills with pollen to carry it back to the hive.

Courtesy of Bob Peterson from North Palm Beach, Florida, Planet Earth— Wikimedia Commons License CC BY SA 2.0

Two types of seed dispersal structures

Left: The feathery structure on a dandelion seed *(Taraxacum sp.)* allows it to float.
Right: Hooks on a greater burdock seed *(Arctium lappa)* can attach it to a passerby.

Photo credits: Dandelion—Didier Descouens, CC BY SA 4.0; Greater burdock—Roger Culos, CC BY SA 3.0 [Both photos: Wikimedia Commons]

Once seeds mature, they are dispersed, or spread, in a number of different ways. Some seeds, such as dandelions, have feathery structures that allow them to be carried by the wind, while maple and elm trees have papery structures for the same purpose. Other seeds have hooks, barbs, spines, or sticky mucus to attach the seeds to birds and other animals for transportation to a new location. Still other seeds can pass through an animal's digestive tract without being damaged. The fruit that covers seeds attracts animals that eat the fruit. Seeds are spread when they pass through an animal's digestive tract and land on the ground along with manure from which a sprouting seed can get nutrients. Water can be used as a way to disperse some seeds, and other seeds form in pods that burst open when ripe, exploding the seeds away from the parent plant.

The word angiosperm comes from two Greek words: *angeion*, meaning "vessel," and *sperma*, meaning "seed." Flowers are the reproductive organs for angiosperms and have several different kinds of specialized leaves called sepals, petals, stamens, and carpels. Angiosperms are those plants with seeds that develop inside a vessel called a carpel. Like gymnosperms, the angiosperm plant that you see

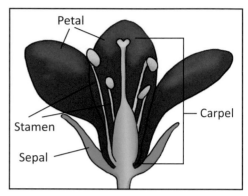

is the sporophyte. The gametophyte is held within the flower. Of all the different types of plants, angiosperm sporophytes provide the most nurturing for the gametophyte.

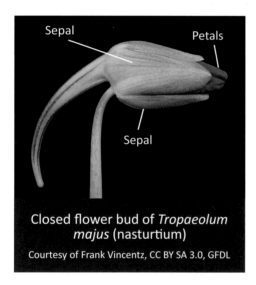

Closed flower bud of *Tropaeolum majus* (nasturtium)

Courtesy of Frank Vincentz, CC BY SA 3.0, GFDL

On the outside of a flower and near the base are the sepals. In many plants, sepals are green and resemble ordinary leaves. Sepals keep a bud closed and protected while it's developing. When the sepal opens, it reveals the petals which surround the reproductive organs of the flower and are typically brightly colored. Petals attract insects such as butterflies and bees that help pollinate the flower. Inside the petals are the stamens and carpel that produce the male and female gametes.

Stamens on the flower of *Hippeastrum sp.* (amaryllis)

Courtesy of André Karwath aka, Wikimedia Commons CC BY SA 2.5

The stamen produces the male gametes and is made of an anther and a filament. The anther is a small sac that houses microsporangia. In a microsporangium a microspore develops into a microscopic pollen grain that contains a male gamete (sperm cell). The anther is held up by the filament, a long thin stalk that sticks up toward the top of the petals. Often there are small structures called nectaries at the base of the stamens. The nectaries produce nectar as the food reward that attracts insects and some birds.

In the very center of the flower is the carpel, which is made up of an ovary, a style, and a stigma. A flower can have one or more carpels depending on the species. The ovary is found at the base of the carpel and contains ridges of tissue that hold one or more ovules where female gametes are produced. From the ovary extends an elongated stalk called the style that supports the stigma. The stigma captures pollen by means of a surface that is sticky or hairy. When a pollen grain lands on the stigma, it forms a pollen tube that grows through the style and the ovary until it reaches an egg cell and fertilization takes place. Once fertilization is complete, the flower supports the growing embryo with nutrients. As the seed matures, the ovary wall of the carpel begins to thicken, enclosing the embryo and forming a fruit.

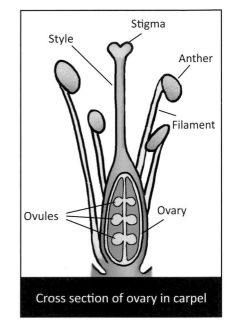

Cross section of ovary in carpel

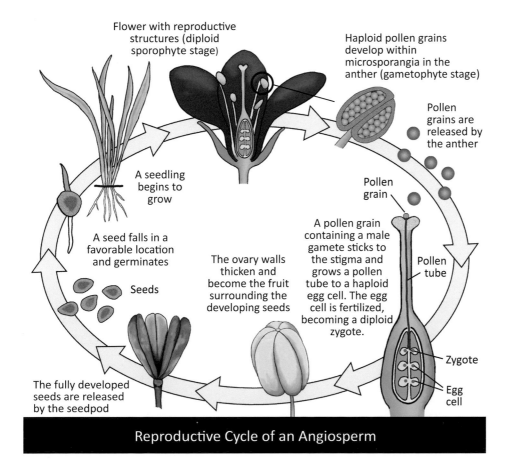

Reproductive Cycle of an Angiosperm

Like other seeded plants, the angiosperm sporophyte is diploid and is the dominant stage. The haploid gametophytes are contained within the flower and are nurtured by the sporophyte. Diploid seeds are created by the gametophyte and develop into new sporophytes.

It is estimated that 80% or more of all plant species are angiosperms. They are adapted to grow in all kinds of climates and environments from very wet to very dry and everything in between . Humans, other animals, birds, and insects all depend on angiosperms for food.

12.4 Summary

- Asexual reproduction in plants is called vegetative propagation. A new plant produced by vegetative propagation is genetically identical to the parent plant.

- Seeded vascular plants reproduce asexually using modified stems, leaves, or roots.

- As a result of sexual reproduction, nonvascular plants and non-seeded vascular plants produce spores that are capable of growing into new plants.

- Vascular seeded plants produce seeds through sexual reproduction. Each seed is capable of growing into a new plant.

12.5 Some Things to Think About

- Why do you think different plants have different methods of reproduction?

- Describe three types of vegetative propagation used by vascular plants.

- What advantages do seeded plants have over non-seeded plants?

- Briefly describe two different plant sexual reproductive cycles.

Chapter 13   Animals

20 μm

13.1 What Is an Animal?

Every day, when you wake up, chances are you will come across an animal of some kind. You might see a spider on the wall of your bedroom as you get out of bed, or you might see an ant running across the bathroom

floor as you are brushing your teeth. Your family might have a pet cat or dog that might greet you with a wet nose or familiar meow as you walk into the kitchen to fetch breakfast. Or you might have a snake or gecko or goldfish in the family den. If you have a sibling or two, you will have greeted other members of the animal kingdom.

Although animals vary widely from one another, there are a few traits that all animals share. All animals are heterotrophs, meaning they don't produce their own food but obtain their nutrients from complex organic compounds—other animals and plants. All animals are multicellular and are composed of eukaryotic cells that contain a nucleus, lack a rigid cell wall, and contain membrane-bound organelles. With the exception of sponges, animal bodies are made up of cells that are organized into tissues that in turn are organized into organs that perform specific functions.

Most animals are mobile for at least some part of their life cycle, and most animals reproduce sexually, with each individual having a unique genetic makeup that is the result of the combining of genetic material contained in an egg and in a sperm. Most animals are non-chordates, meaning they do not have a backbone, or vertebral column. Non-chordates include sponges, sea stars, worms, jellyfish, and insects. Only 5% of animals are chordates with a backbone or vertebral column. Chordates include many large animals like whales and elephants and small animals like fish, birds, and reptiles.

The exact number of animals that inhabit Earth is not known, but scientists estimate that there are somewhere between 8 and 10 million animal species. Scientists estimate that 6.5 million species live on land with about 2.2 million species living in the oceans (give or take 1.3 million)[1].

Because Earth is so large with parts of the deep ocean and land masses that have not been explored, it's difficult to figure out exactly how many animals inhabit our planet. Also, because there are so many different kinds of animals, it is difficult to keep track of all of the animals we have discovered and equally challenging to organize them into categories. Genetic tools are helping scientists catalog and organize many animals, but even genetic classification methods can get complicated quickly. Taxonomic categories are undergoing changes as more data is collected and new ideas are introduced.

☐ **Invertebrates—95%**   ☐ **Vertebrates—5%**

13.2 Animal Cells

One trait all animals share, whether they live on land or in the ocean, fly in the sky, swim in the rivers, or drive cars, is that all animals are made of animal cells. An animal cell is a type of eukaryotic cell. Unlike a bacterial cell or archaeal cell, an animal cell has a membrane-bound nucleus that contains the genetic information encoded in DNA.

Animal cells are surrounded by a plasma membrane and filled with a water-based fluid called cytosol. Suspended in the cytosol are structures called organelles that perform specialized functions. Also found in the cytosol is the cytoskeleton, an elaborate array of specialized proteins that serves as a sort of superhighway for molecular transport. Together, the cytosol and organelles make up the cellular cytoplasm.

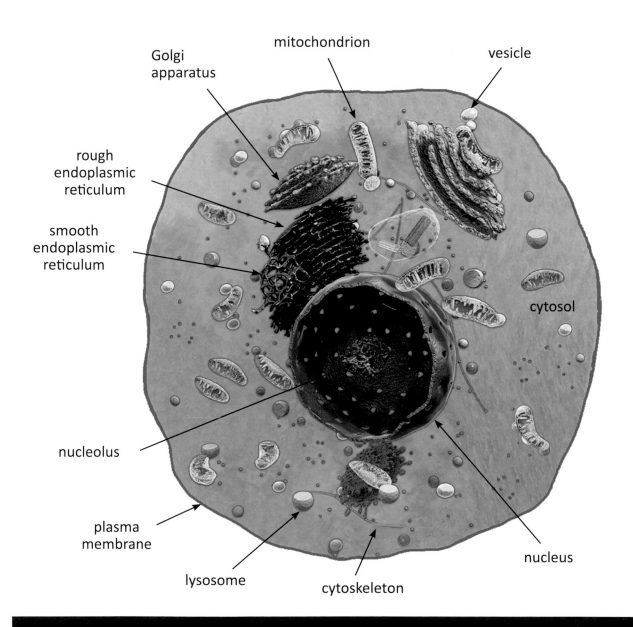

Animal Cell

Adapted from illustration by Blausen.com staff (2014). "Medical gallery of Blausen Medical 2014". WikiJournal of Medicine 1 (2). DOI:10.15347/wjm/2014.010. ISSN 2002-4436.

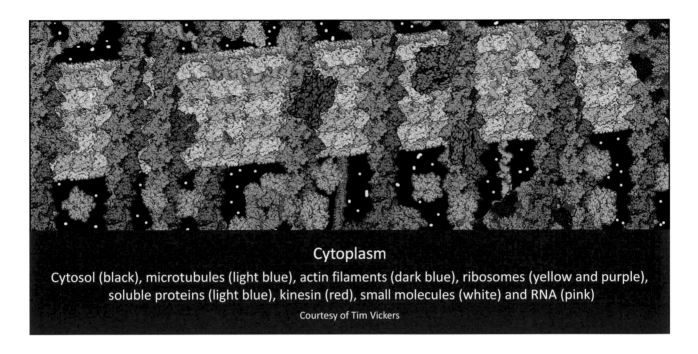

Cytoplasm
Cytosol (black), microtubules (light blue), actin filaments (dark blue), ribosomes (yellow and purple), soluble proteins (light blue), kinesin (red), small molecules (white) and RNA (pink)
Courtesy of Tim Vickers

Parts of a Cell —A Closer Look

The Plasma Membrane

All cells, including both prokaryotic and eukaryotic cells, are surrounded by a plasma membrane. Some organelles are also surrounded by a plasma membrane. The plasma membrane defines the boundary of the cell or organelle and separates it from its surroundings. A plasma membrane is a semipermeable barrier, meaning that it can allow some molecules through while excluding others.

The plasma membrane is made of lipids and proteins that together form a lipid bilayer. A lipid bilayer is made of two layers of lipids and proteins that are pushed together with the lipids occupying the inside of the bilayer and the proteins occupying the outside of the bilayer.

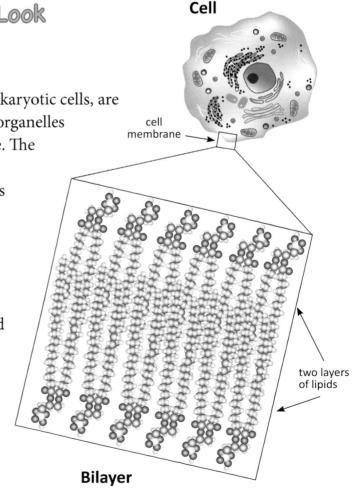

Cell

cell membrane

two layers of lipids

Bilayer

Organelles

In order to fully understand how an animal cell works, it is essential to know how each of the organelles inside the cell works, how molecules are moved between organelles, and how the organelles are created and maintained. Each organelle has its own set of proteins and other specialized molecules and performs a particular function. Let's take a closer look at some of the organelles found in animal cells. Table 13.1 lists some of the most common ones.

Nucleus

An animal cell contains a nucleus that holds and processes the cell's genetic information. DNA, RNA, and a host of enzymes work together to replicate, modify, and regulate genetic information.

The nucleus is surrounded by two membranes referred to as the nuclear envelope. These inner and outer membranes together separate the contents of the nucleus from the cytoplasm. The cytoplasm includes the entire contents of a cell outside

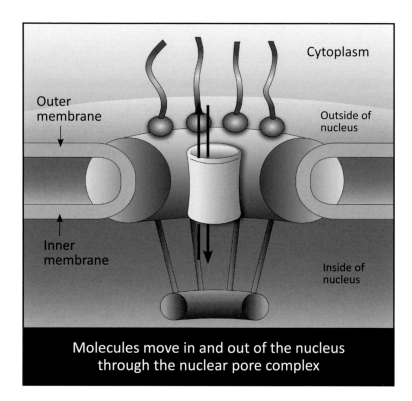

Molecules move in and out of the nucleus through the nuclear pore complex

the nucleus. The two membranes of the nuclear envelope act as a barrier that prevents molecules from flowing freely between the nucleus and the cytoplasm. However, the nuclear membranes contain channels called nuclear pore complexes that act as little gatekeepers, moving some molecules through the membrane and not allowing others to pass.

The most important aspect of the nucleus is the storage and duplication of the cell's chromosomes. Chromosomes are long wound up strands of DNA and protein that are pulled apart for duplication during cell division. Animal cells have multiple chromosomes that sit inside the nucleus as tightly packed bundles waiting for the cell to send them a signal to divide. When the cell gives the go-ahead for cell division, proteins go to work pulling the strands of DNA, copying that DNA, and then moving the two copies into separate sections of the cell so that when the cell divides, each new cell has its own DNA copy.

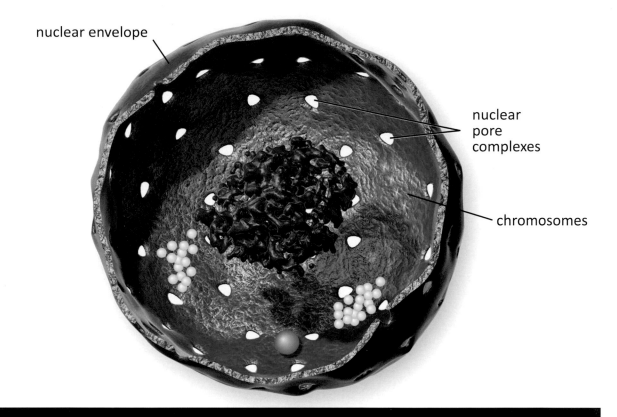

nuclear envelope

nuclear pore complexes

chromosomes

Nucleus

Adapted from Illustration by Blausen.com staff (2014). "Medical gallery of Blausen Medical 2014". WikiJournal of Medicine 1 (2). DOI:10.15347/wjm/2014.010. ISSN 2002-4436.

Endoplasmic Reticulum

The endoplasmic reticulum (ER) is attached to the outside of the nuclear membrane and extends into the cell cytosol. The ER is a network of sacs and tubules that are enclosed by a continuous membrane and is the largest organelle inside the cytosol of most eukaryotic cells.

There are two different domains, or regions, in the ER that perform different functions. These are the rough ER and the smooth ER.

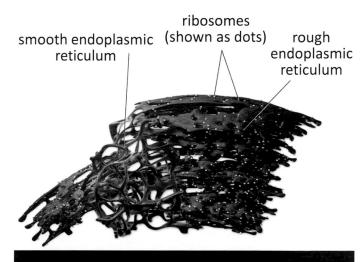

smooth endoplasmic reticulum

ribosomes (shown as dots)

rough endoplasmic reticulum

Endoplasmic reticulum

Courtesy of Blausen.com staff (2014). "Medical gallery of Blausen Medical 2014". WikiJournal of Medicine 1 (2). DOI:10.15347/wjm/2014.010. ISSN 2002-4436.

The rough ER is dotted with ribosomes that cover the outer surface of the membrane that is in contact with the cytosol. This outer membrane surface is called the cytosolic surface. The rough ER is involved in protein synthesis, the building of proteins.

The smooth ER is not dotted with ribosomes and is involved in manufacturing lipids rather than protein synthesis.

Golgi Apparatus

The Golgi apparatus (also called the Golgi complex), is a sort of post office for proteins. Proteins that have been synthesized come off the ER and go to the Golgi apparatus for further processing, sorting, and eventual transportation to their destinations.

The Golgi is made of membrane-bound, flattened and stacked sacs and smaller membrane-bound spherical sacs (called vesicles)

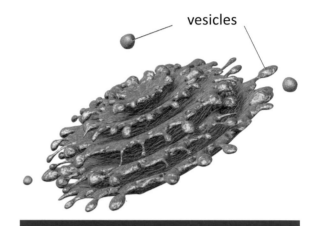

vesicles

Golgi apparatus
Courtesy of Blausen.com staff (2014). "Medical gallery of Blausen Medical 2014". WikiJournal of Medicine 1 (2). DOI:10.15347/wjm/2014.010. ISSN 2002-4436.

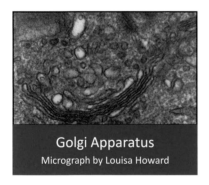

Golgi Apparatus
Micrograph by Louisa Howard

that together transport and process proteins. Proteins coming off the ER are carried by vesicles that merge with one side of the Golgi. Once merged, proteins move through the flattened and stacked sacs for processing and sorting. After the proteins are processed and sorted, they emerge from the Golgi in another vesicle for transportation to other areas of the cell.

Lysosomes

Lysosomes are like little recycling factories for dead, damaged, or used proteins. Lysosomes contain enzymes that break down proteins the cell no longer needs. Lysosomes break

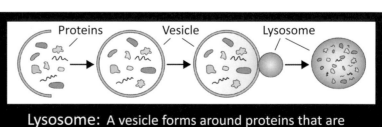

Proteins   Vesicle   Lysosome

Lysosome: A vesicle forms around proteins that are no longer needed and merges with a lysosome where the proteins are broken down and then recycled.

up cellular proteins by a process called autophagy. [Autophagy comes from the Greek words *autos,* meaning "self"; and *phagein,* "to eat." So autophagy refers to the process of a cell consuming part of itself.] When a protein is no longer needed, it is surrounded by plasma membrane, often pinched off from the ER. A vesicle, or small spherical sac, is formed that has plasma membrane surrounding the protein. A lysosome merges with the vesicle, forming a larger vesicle where the protein can be broken down and eventually recycled.

Mitochondria

Mitochondria are the powerhouse organelles of the cell. Mitochondria generate energy by breaking down carbohydrates and fatty acids and converting them to ATP (adenosine triphosphate). ATP is the primary energy molecule used by the cell.

A mitochondrion has an inner and outer membrane with a space in between them. The inner membrane forms numerous folds called cristae that extend towards the center of the mitochondrion. The inner membrane houses the sophisticated machinery that produces ATP.

inner membrane

outer membrane

cristae

Mitochondrion

Courtesy of Blausen.com staff (2014). "Medical gallery of Blausen Medical 2014". WikiJournal of Medicine 1 (2). DOI:10.15347/wjm/2014.010. ISSN 2002-4436.

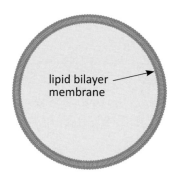

lipid bilayer membrane

Peroxisomes

A peroxisome is a small organelle with a single lipid bilayer membrane enclosing it. Peroxisomes contain enzymes that are involved in a number of metabolic reactions, including the breakdown of hydrogen peroxide and making lipids and certain steroids such as cholesterol.

The Cytoskeleton

The cytoskeleton acts like a cellular superhighway. Proteins, molecules, and vesicles are transported throughout the cell on the cytoskeleton.

The cytoskeleton is made up of a protein called actin that forms long filaments (microfilaments). Actin microfilaments are thin and flexible and can be easily assembled

Table 13.1 Some Animal Cell Organelles

Cytoskeleton		A structure made of long protein filaments; acts like a cellular superhighway for transporting proteins, molecules, and vesicles throughout the cell
Golgi apparatus		Takes in proteins that have been synthesized in the endoplasmic reticulum; processes and sorts the proteins to ready them for transport by vesicles
Lysosome		Contains enzymes that break down proteins that are no longer needed by the cell
Mitochondrion		Generates energy by breaking down carbohydrates and fatty acids and converting them to ATP (adenosine triphosphate)
Nucleus		Holds and processes the cell's genetic information; DNA, RNA, and enzymes work together in the nucleus to replicate, modify, and regulate genetic information
Peroxisome		Contains enzymes involved in a number of metabolic reactions, including the breakdown of hydrogen peroxide and making lipids and certain steroids such as cholesterol
Rough Endoplasmic Reticulum		A network of sacs and tubules that has ribosomes on its surface; involved in building proteins (protein synthesis)
Smooth Endoplasmic Reticulum		A network of sacs and tubules that is involved in lipid metabolism

and disassembled, branching from one end of the cell to the other and creating a sophisticated transportation network.

Actin was initially discovered in muscle tissue and thought to be only involved in muscle contraction. However, we now know that actin is found in all types of eukaryotic cells. Mammals have different types of actin found in skeletal, cardiac, and smooth muscle tissue, and two other types of actin are found in non-muscle tissue.

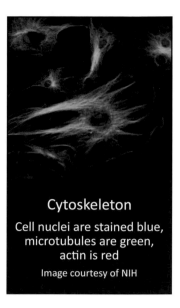

Cytoskeleton
Cell nuclei are stained blue, microtubules are green, actin is red
Image courtesy of NIH

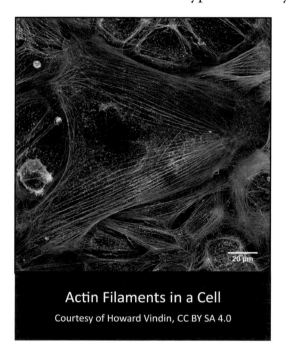

Actin Filaments in a Cell
Courtesy of Howard Vindin, CC BY SA 4.0

Actin microfilaments can be organized into ladder-style parallel packed bundles or in a crisscrossed linked network. Actin bundles and actin networks allow transport of molecules inside the cell and contraction of muscle tissues, and they support structures for cell formation and cell shape.

13.3 Different Types of Animal Cells

Although all animal cells have a similar basic structure, there are many different types of animal cells in any one organism and between different types of organisms. During development, animal cells differentiate, or change, as they mature. As cells develop, they can become specialized and can combine to form specialized tissues and specific organs.

Having specialized cells means that animals can perform a variety of functions more efficiently than single-celled organisms. When specialized cells group together to form specialized tissues and organs, animals gain an advantage when regulating body temperature, digesting a variety of foods, running, jumping, and reproducing.

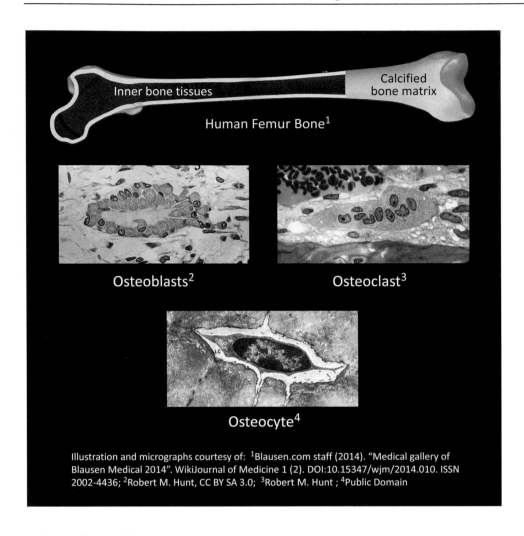

Inner bone tissues

Calcified bone matrix

Human Femur Bone[1]

Osteoblasts[2]

Osteoclast[3]

Osteocyte[4]

Illustration and micrographs courtesy of: [1]Blausen.com staff (2014). "Medical gallery of Blausen Medical 2014". WikiJournal of Medicine 1 (2). DOI:10.15347/wjm/2014.010. ISSN 2002-4436; [2]Robert M. Hunt, CC BY SA 3.0; [3]Robert M. Hunt ; [4]Public Domain

Bone Cells

Animals that have internal skeletons are supported by bones. Bones are made of three different types of bone cells called osteocytes, osteoblasts, and osteoclasts, which together form the soft inner tissues and a hardened calcified matrix.

Blood Cells

There are two different types of blood cells: red blood cells and white blood cells. Red blood cells make up 99.9% of all the blood cells in the body and have the job of delivering all the oxygen from the lungs to the rest of body. Mature red blood cells are the only animal cells that do not have a nucleus, and these cells are produced in the bone marrow found in the interior of bones. Initially, red blood cells do have a nucleus, but they undergo a process

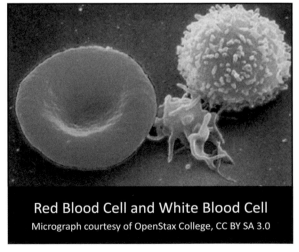

Red Blood Cell and White Blood Cell
Micrograph courtesy of OpenStax College, CC BY SA 3.0

called enucleation in which the nucleus is pulled out of the cell. Lacking a nucleus allows red blood cells to carry more oxygen and to move more freely through the body.

White blood cells are part of the immune system, which protects the body from foreign substances such as harmful bacteria. There are several different types of white blood cells whose job it is to battle unwanted bacteria, viruses, and other pathogens that could do harm to the body. White blood cells are produced in the bone marrow, as are red blood cells, but white blood cells do not undergo enucleation; therefore, they do have a nucleus.

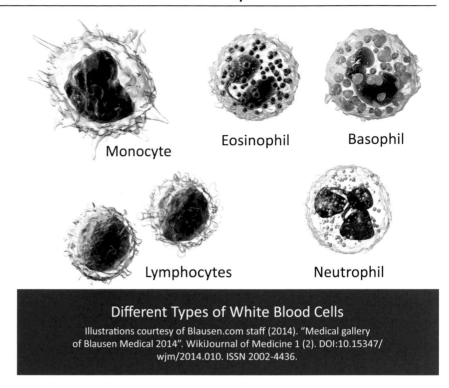

Monocyte   Eosinophil   Basophil

Lymphocytes   Neutrophil

Different Types of White Blood Cells

Illustrations courtesy of Blausen.com staff (2014). "Medical gallery of Blausen Medical 2014". WikiJournal of Medicine 1 (2). DOI:10.15347/wjm/2014.010. ISSN 2002-4436.

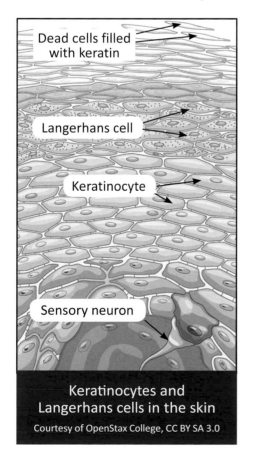

Dead cells filled with keratin

Langerhans cell

Keratinocyte

Sensory neuron

Keratinocytes and Langerhans cells in the skin

Courtesy of OpenStax College, CC BY SA 3.0

Skin Cells

The skin is the largest and one of the most important organs of the human body. Skin cells in animals are made mainly of three types of cells—keratinocytes, melanocytes, and Langerhans cells.

Keratinocytes are the most abundant cell type and produce a protein called keratin. Keratin forms long filaments that become hair, nails, and hooves. Keratin also forms a protective barrier on the surface of the skin.

Melanocytes are the second most abundant type of skin cell. They produce a compound called melanin, which gives skin its color.

Langerhans cells are in the outer layer of skin and help protect the body from microbial invaders.

Muscle Cells

When you flex your arm or jump over a rock, your muscle cells leap into action. There are several different types of muscle cells.

Muscle cells are classified as either striated or smooth according to their appearance. Striated muscle cells have cross-striations, which look like little lines that go horizontally across a muscle cell or fiber. These cross-striations are made of overlapping thick and thin protein strands called myofilaments. Smooth muscle cells do not have cross-striations.

Striated muscle cells form skeletal muscle, which is responsible for making movements of the body, and cardiac muscle, which controls contractions in the heart. Smooth muscle cells control subconscious movements of tissues, such as those in blood vessels, the stomach, and some of the muscles of the eye.

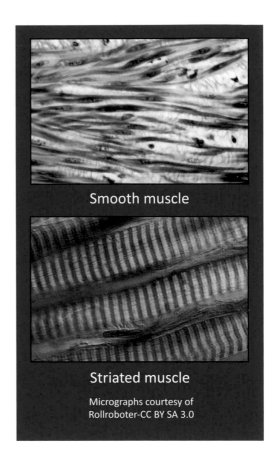

Smooth muscle

Striated muscle

Micrographs courtesy of
Rollroboter-CC BY SA 3.0

Nerve Cells

Nerve cells, also called neurons, are cells that transmit electrical or chemical signals from one cell to another, relaying information about inputs such as pressure, temperature, light, and sound. Neurons are the main cells of the nervous system, which is a network of cells that moves signals throughout the body.

In general, neurons receive stimuli from other cells and send electrical impulses to different parts of the nervous system. They are oddly shaped cells with a centralized cell body containing the nucleus and a long process (projecting part) called an axon that extends outward from the cell body. Neurons also have shorter processes called dendrites that extend radially from the cell body and receive signals from the surrounding cells. Neurons have specialized structures called synapses that contain molecular machinery that allows information to be passed from one neuron to another. The human brain has around 100 billion neurons!

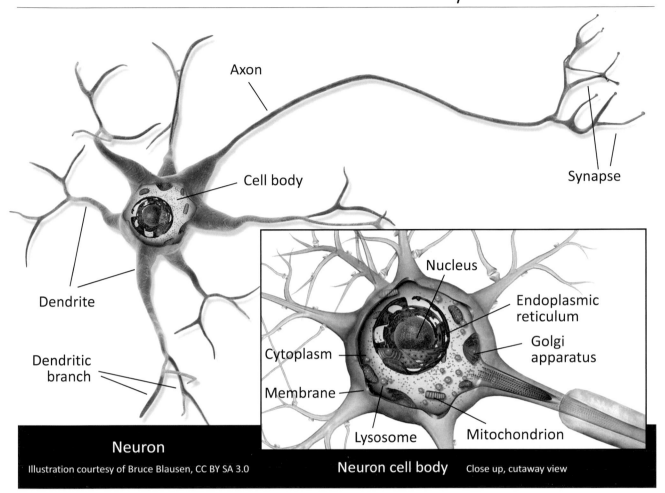

Neuron
Illustration courtesy of Bruce Blausen, CC BY SA 3.0

Neuron cell body Close up, cutaway view

13.4 Animal Phyla

Within the animal kingdom and under the domain Eukarya, there are many different animal phyla. Recall that in taxonomy a phylum is a subdivision of a kingdom.

Scientists are still arguing about how to classify all animals according to phyla, and as new discoveries are made, some of these divisions may change. Historically, animals have been divided into two broad categories: non-chordates and chordates.

Chordates have a notochord as they begin developing. A notochord is a skeletal rod made mostly of cartilage-like material that runs along the back of the animal. The notochord contains a column of nerve cells and connects the brain and the rest of the body. For animals that are vertebrates, the notochord develops into a spinal cord protected by a rigid column of bones called vertebrae. Non-chordates do not have a notochord, spinal cord, or vertebral column. We will learn more about chordates, non-chordates, and animal phyla in the remaining chapters.

Animal Phyla

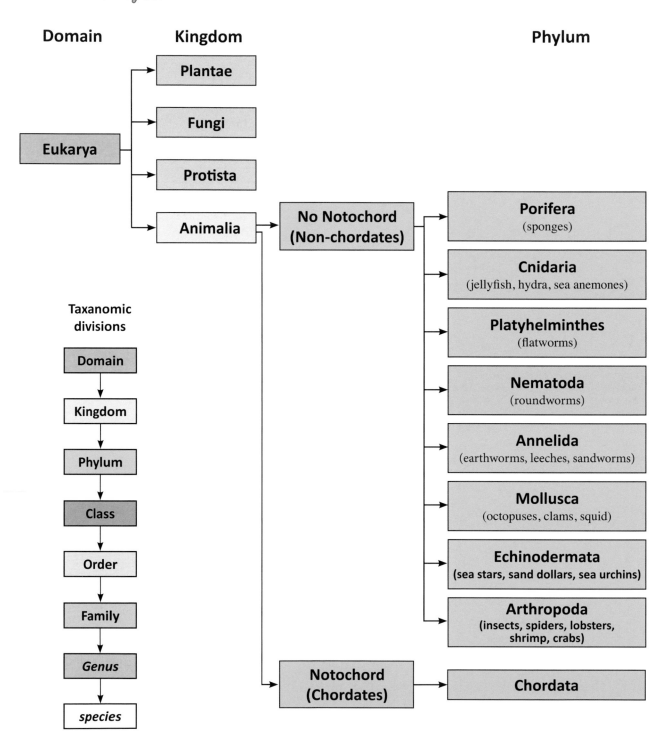

Domain

Kingdom

Phylum

Eukarya

Plantae

Fungi

Protista

Animalia

No Notochord
(Non-chordates)

Notochord
(Chordates)

Porifera
(sponges)

Cnidaria
(jellyfish, hydra, sea anemones)

Platyhelminthes
(flatworms)

Nematoda
(roundworms)

Annelida
(earthworms, leeches, sandworms)

Mollusca
(octopuses, clams, squid)

Echinodermata
(sea stars, sand dollars, sea urchins)

Arthropoda
(insects, spiders, lobsters,
shrimp, crabs)

Chordata

Taxanomic
divisions

Domain

Kingdom

Phylum

Class

Order

Family

Genus

species

13.5 Summary

- Animals are heterotrophs. They cannot make their own food and need to obtain nutrients by eating plants and/or animals.

- It is estimated that there are 8-10 million different animal species.

- All animals have animal cells with a plasma membrane and organelles.

- The plasma membrane surrounding an animal cell is a lipid bilayer and is semipermeable.

- An organelle is a small structure that performs a specialized function inside a cell.

- Many animals have specialized cells that perform specialized functions.

- In taxonomy animals are grouped into chordates and non-chordates according to whether or not the animal has a notochord as some point in its life.

13.6 Some Things to Think About

- Why do you think most animals are non-chordates?

- Look up zip code proteins on the internet or at the library. Why do you think proteins have zip codes?

- Why do you think red blood cells and white blood cells look so different?

- Look at the shape of a neuron. Why do you think it is shaped this way?

References

1. Camilo Mora, Derek P. Tittensor, Sina Adl, Alastair G. B. Simpson, Boris Worm (2011); "How Many Species Are There on Earth and in the Ocean?" *PLoS Biology* 9(8): e1001127. https://doi.org/10.1371/journal.pbio.1001127

Chapter 14 Non-chordates

14.1 Introduction

As we discovered in the last chapter, animals can be broadly divided into two main categories: non-chordates and chordates. In this chapter, we will take a closer look at the main phyla that make up non-chordates. There are eight phyla in the kingdom Animalia that are non-chordates, including sponges, jellyfish, worms, mollusks, and arthropods. We will take a closer look at chordates in the next chapter.

Non-chordate phyla
Porifera
Cnidaria
Platyhelminthes
Nematoda
Annelida
Mollusca
Echinodermata
Arthropoda

14.2 Phylum Porifera

Sponges are members of the phylum Porifera. The name Porifera comes from two Latin words: *porus* which means "pore" and *ferre,* "to bear." Porifiera means to bear pores, or pore-bearers.

Unlike other animals, sponges have no body symmetry. They are asymmetric with no front, back, left, or right side.

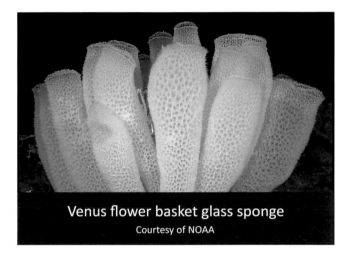

Venus flower basket glass sponge
Courtesy of NOAA

Sponges are the only animals that are asymmetric. Although sponges don't look like other animals, they are classified as animals because they are multicellular, have eukaryotic cells with no cell walls, and are heterotrophic (they cannot manufacture their own food but must eat other organisms). Like other animals, sponges also have specialized cells. Recall from the previous chapter that having specialized cells is advantageous for organisms because different cell types carry out different functions and can form different specialized tissues and organs. Cells that are all of the same type cannot do this.

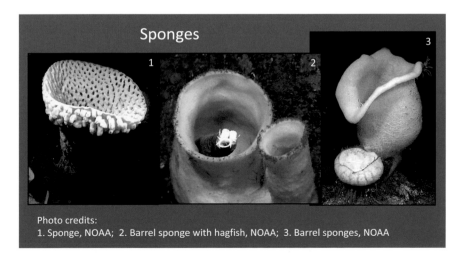

Sponges

Photo credits:
1. Sponge, NOAA; 2. Barrel sponge with hagfish, NOAA; 3. Barrel sponges, NOAA

Sponges usually live attached to the sea floor several meters or more below the water surface. At this depth, there is very little sunlight. Scientists have wondered how some organisms that rely on photosynthesis, such as algae, can live inside sponges where there is so little sunlight. They've discovered that the cross-shaped spicules of some sponges are like little antennae for sunlight and act like tiny magnifying glasses that focus sunlight, allowing photosynthetic organisms to live and grow[1].

14.3 Phylum Cnidaria

The phylum Cnidaria contains a number of soft-bodied animals, including jellyfish, hydras, sea anemones, and corals. These animals are carnivorous (eat other animals) and have specialized tissues. Cnidarians get their name from stinging cells called cnidocytes located in their tentacles. A cnidocyte contains an organelle called a nematocyst that produces poison and contains a tightly coiled dart that is flung out by the cnidarian to capture prey by impaling and poisoning it. The nematocyst is also used for self-defense. The name Cnidaria comes from the Greek word *cnidos,* which means "stinging nettle."

All Cnidarians have radial body symmetry. If you take an imaginary plane dividing any two halves along the length of the body you would see a mirror image. The tentacles radiate outward like spokes on a wheel, giving Cnidarians their radial symmetry.

Radial symmetry

Photo credits: 1. Jellyfish, Francesco Crippa, CC BY SA 2.0; 2. Sea anemone, NOAA Okeanos Explorer Program, Galapagos Rift Expedition 2011; 3. Coral reef, Toby Hudson, CC BY SA 3.0

Cnidarians paralyze their prey with a stinging nematocyst. After the prey is paralyzed, the cnidarian uses its tentacles to pull the prey toward the cnidarian's central cavity, which is lined with gastrodermal tissue. As the food digests, it is absorbed through the cells in the gastrodermal layer, and anything that is not absorbed is pushed out of the central cavity opening.

Cnidarians have two body forms: the polyp and the medusa. Polyps are generally immobile and attach to a surface, with the mouth and tentacles pointing upward. The medusa stage is mobile and free-floating. The body is bell-shaped with the mouth and tentacles pointing downward. Some cnidarians, such as sea anemones and hydras, are always polyps, some are always medusae, like some jellyfish, and others, such as hydrozoa, have both a polyp and a medusa stage.

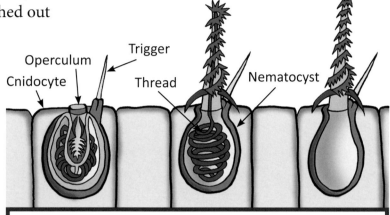

A nematocyst firing

Left: The nematocyst is inside its capsule within the cnidocyte of the tentacle. The thread of the nematocyst is coiled and under pressure.

Middle: When prey makes contact with the trigger mechanism, the nematocyst is activated. The operculum, the flap of tissue covering the nematocyst, flies open, and the dart covered with barbs is forced out by the pressure within the capsule.

Right: The dart sticks into the prey and poison is ejected. When the prey is subdued by the poison, the cnidarian can use its tentacles to draw the food toward its mouth. The thread attached to the dart allows it to be drawn back into the capsule once it has been released from the prey.

14.4 Worms

Worm phyla

| Platyhelminthes |
| Nematoda |
| Annelida |

Worms! Some worms can help grow a garden, and other worms cause problems for people. Some worms float beautifully under the water, and some have spines and pinchers, looking like something from outer space. Some worms are so small that you need a microscope to see them. Other worms are huge, like the Giant Gippsland earthworm in Australia, which can grow to be 3 meters (almost 10') long! Worms can be long and thin or short and fat, and they move by undulating their bodies, whether on a surface or in a liquid.

There are three main types of worms: flatworms, roundworms, and annelids. Flatworms belong to the phylum Platyhelminthes. Roundworms belong to the phylum Nematoda, and annelids belong to the phylum Annelida.

Phylum Platyhelminthes

Like the name implies, flatworms have a flattened and spread out body that is no more than a few millimeters thick. Most flatworms do not have a coelom, which is an internal fluid-filled body cavity. For this reason they are called acoelomates. Recall that the prefix "a-" means without, so the word acoelomate means "without coelom."

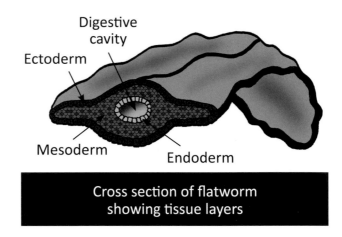

Cross section of flatworm showing tissue layers

Flatworms have specialized cells and tissues, including an ectoderm (outer tissue layer), a mesoderm (middle tissue layer), and an endoderm (inner tissue layer). The endoderm serves as the digestive tissue. Because flatworms have no coelom, in larger flatworms the digestive tissue, or gut, is often highly branched and extends to all parts of the body.

In addition to specialized cells and tissues, most flatworms share some basic body structures, including an eyespot, head, mouth, and a nervous system. Flatworms don't have a circulatory system and move oxygen and nutrients in and out of their body mainly through diffusion. Some flatworms have specialized cells called flame cells, that remove wastes and excess water.

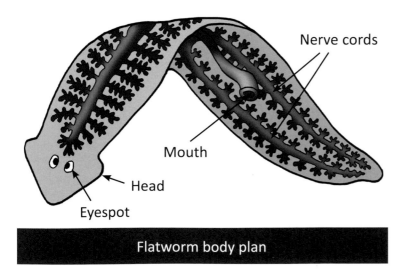

Flatworm body plan

There is a huge variety of flatworms found in freshwater and ocean marine environments and in moist soil. Flatworms can be beautiful with multicolored ribbon-like edges or so tiny you can barely see them with an unaided eye. Some flatworms are parasitic and cause diseases in other animals.

Phylum Nematoda

Unlike flatworms, roundworms have an internal cavity between the endoderm and mesoderm, and they have a rounded body. Their internal cavity is only partially layered with mesoderm tissue; therefore, the internal cavity is called a pseudocoelom or "false coelom."

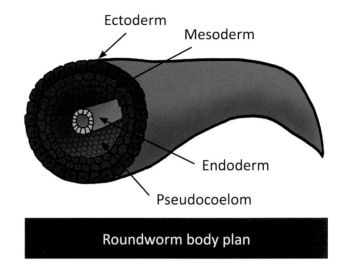

Roundworm body plan

Roundworms come in a variety of sizes but in general are slender and unsegmented, with tapering ends and a head and tail. Unlike flatworms, roundworms have an internal digestive cavity with a mouth and anus, allowing digestion to move in one direction along the body.

Some roundworms are free-living, meaning they do not depend on a host for survival. Free-living roundworms live in soil and in ocean or freshwater environments and eat other small animals, algae, or decaying plants and animals.

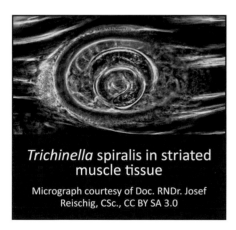

Trichinella spiralis in striated muscle tissue

Micrograph courtesy of Doc. RNDr. Josef Reischig, CSc., CC BY SA 3.0

Some roundworms are parasitic and live inside a host animal. Many roundworms cause disease in humans. Trichinosis is a disease caused by the roundworm *Trichinella spiralis*. *Trichinella spiralis* worms are found in pigs or wild carnivores, such as bears. People get trichinosis when they eat meat from these animals that has been undercooked.

Hookworms are another common disease-causing roundworm. There are two species of hookworms that infect humans—*Ancylostoma duodenale* and *Necator americanus*. Hookworms infect humans by attaching to and penetrating the skin and can cause itching, rash, abdominal pain, and weight loss. Hookworms are often found in areas that do not have adequate sanitation.

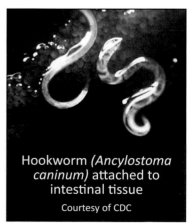

Hookworm *(Ancylostoma caninum)* attached to intestinal tissue

Courtesy of CDC

Phylum Annelida

If you have ever grown a garden, you might have noticed little worms wriggling in the soil. These worms are called earthworms, and they perform the very valuable job of processing decaying organic matter in the soil. They keep soil healthy and full of nutrients so vegetables, flowers, and herbs can grow.

An earthworm
Courtesy of Rob Hille, CC BY SA 3.0

Earthworms are a member of the phylum Annelida. Annelids include leeches, earthworms, and a huge variety of marine worms, such as tube worms and feather dusters. The word Annelida comes from the Latin word *annellus* which means little ring. The bodies of annelids are segmented, which gives them their name. Segments or groups of segments can act independently, and some segments may carry one or more pairs of eyes, antennae, and other organs. Some segments may also be specialized and perform particular functions, such as respiration or reproduction.

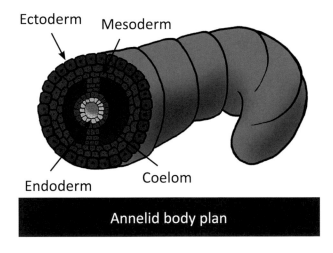

Annelid body plan

Ectoderm Mesoderm

Endoderm Coelom

Annelids have a true coelom lined with mesodermal tissue. Like roundworms, annelids have a mouth and anus allowing digestion to move in one direction along the body. Unlike roundworms and flatworms, annelids have a closed circulatory system. Blood cells circulate through a network of blood vessels, allowing oxygen and nutrients to be distributed throughout the body.

14.5 Phylum Mollusca

If you've ever had the chance to eat in a French restaurant, you might have noticed a few mollusks on the menu. Many cultures eat a variety of mollusks, including clams, snails, squids, and octopuses. Mollusks are the second most abundant animal phylum on the planet.

Mollusks are in the phylum Mollusca. The word mollusk comes from the Latin word *molluscus* which means "soft." Mollusks are soft-bodied animals with an internal or external shell. Mollusks are a diverse group of animals with nearly 150,000 different species. Some are so tiny, you can barely see them with the unaided eye, and other mollusks, such as the giant squid, can be 20 meters (65 feet) long.

Mollusks can be found in a variety of habitats from soils to oceans. Although mollusks can look very different from one another, they all share a few characteristics. All mollusks have variations on a common body plan, including a soft body with a "head" and a "foot" region, a thin layer of tissue (called the mantle) covering most of the body, and a shell made of calcium carbonate. In some species, such as snails and clams, the shell is obvious,

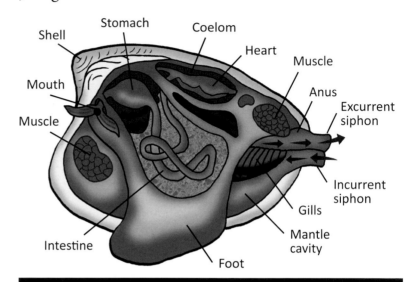

Mollusk basic body plan

The incurrent siphon brings water into the animal. The gills take oxygen from the water. The deoxygenated water then travels to the mouth for food to be filtered out. Water and wastes are expelled through the excurrent siphon.

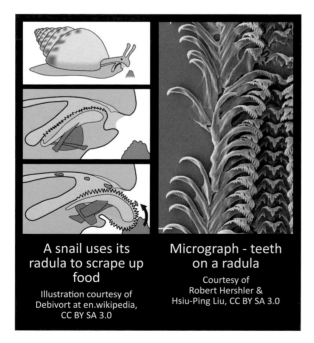

A snail uses its radula to scrape up food

Illustration courtesy of Debivort at en.wikipedia, CC BY SA 3.0

Micrograph - teeth on a radula

Courtesy of Robert Hershler & Hsiu-Ping Liu, CC BY SA 3.0

and in other species, such as octopuses and slugs, the shell has been reduced or lost. Mollusks also have specialized tissues and organs, including a stomach, heart, intestines, muscles, a mouth, and an anus.

Mollusks eat a variety of foods and can be carnivores, herbivores, or filter feeders. Some mollusks, such as snails, have a tongue-like structure called a radula that is covered with rows of small teeth. The radula is used to scrape algae from rocks and other surfaces.

Mollusks found in oceans and freshwater environments use gills to breathe. The gills are located inside the mantle. As water passes over the mantle, oxygen is absorbed by the gills. Blood pumped by a simple heart transports oxygen and other nutrients throughout the mollusk body by means of a circulatory network of vessels and spaces in the body. Snails and clams have this type of structure.

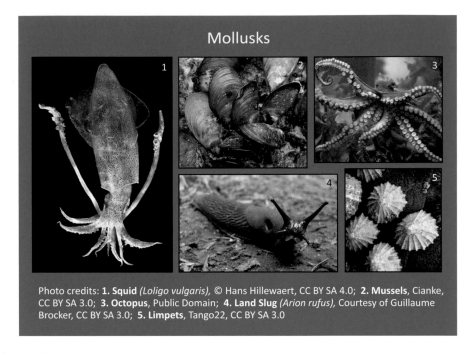

Mollusks

Photo credits: **1. Squid** *(Loligo vulgaris),* © Hans Hillewaert, CC BY SA 4.0; **2. Mussels**, Cianke, CC BY SA 3.0; **3. Octopus**, Public Domain; **4. Land Slug** *(Arion rufus),* Courtesy of Guillaume Brocker, CC BY SA 3.0; **5. Limpets**, Tango22, CC BY SA 3.0

14.6 Phylum Arthropoda

Arthropods are the largest group of animals on the planet. You can find arthropods everywhere. They live in oceans, rivers, and on land, and some spend most of their time flying around in the air, helping flowers bloom or pestering us at picnics.

The word arthropod comes from two Greek words: *arthron* which means "joint" and *podos* which means "foot." Arthropods are animals with jointed legs and antennae and segmented bodies, and many have tough external skeletons made of chitin. Arthropods have specialized tissues and organs to carry out different essential functions. Arthropods have eyes, a brain, a heart with a circulatory system, mouth, anus, a digestive tract with a stomach, and a well-developed nervous system.

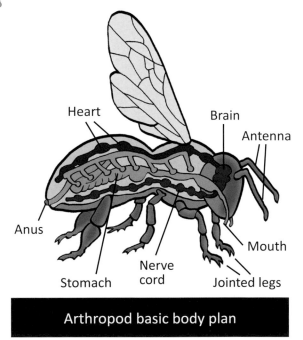

Arthropod basic body plan

Mouthparts

Photo credits: 1. Micrograph of butterfly tongue, Public Domain; 2. Beetle, Siga, CC BY SA 3.0

As a group, arthropods eat just about everything because the phylum Arthropoda includes omnivores, herbivores, and carnivores. Arthropods have specialized mouthparts that enable them to cut tissues, using jaws equipped with pincers or fangs that slice, pinch, or piece prey. Some arthropods have a tough prong that they use to bore into the hard outer shell of nuts.

Arthropods have specialized tissues and organs. Depending on the species, they can utilize lungs for breathing on land or gills for breathing under water. Arthropods move their limbs with well-developed groups of muscles that are controlled by a fully functioning nervous system, and throughout the body they have nerves that are attached to a centralized brain.

Molting

Dragonfly, Courtesy of Mathiasrex, CC BY SA 3.0

Arthropods do not have an internal skeleton but instead have an external covering or exoskeleton. The exoskeleton acts as a container for the organs and for the circulatory, nervous, and reproductive systems. It also provides armor to protect the animal from predators. Since the exoskeleton doesn't grow larger, it must be shed at various times during an arthropod's life in order to accommodate the growing body. The process of shedding the exoskeleton is called molting. Molting can take several hours as the old exoskeleton is removed and replaced with the new, larger exoskeleton, which starts out soft and takes some time to harden.

Because there is such a wide variety of arthropods, scientists have had a difficult time grouping them into appropriate categories. In general, arthropods are classified according to the number of body segments they have and the structure of their body segments and mouthparts. With this in mind, arthropods are divided into three subphyla: Crustacea, Chelicerata, and Uniramia, which is further divided into three classes; Diplopoda, Chilopoda, and Insecta.

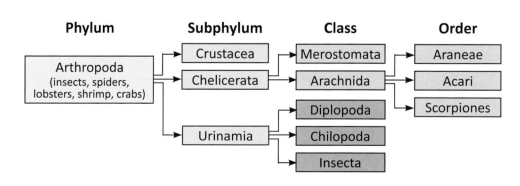

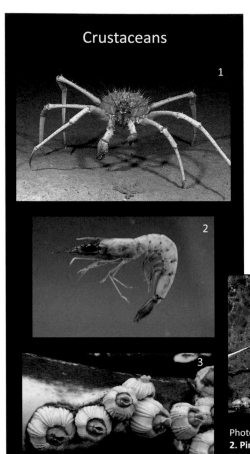

Crustaceans

Subphylum Crustacea

Crustaceans include crabs, shrimp, lobsters, and crayfish and live primarily in water environments, on the ocean floor, in rivers and ponds, and stuck to the sides of boats or other objects found in oceans, lakes, or ponds. In general, crustaceans have two or three body sections, chewing mouthparts called mandibles, and two segmented antennae. Some crustaceans, such as barnacles, have lost their mandibles and abdominal segments.

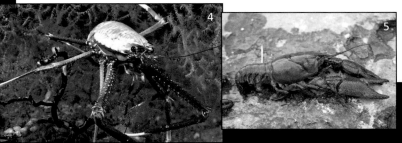

Photo credits: **1. Lithoid crab**, NOAA Okeanos Explorer Program, Gulf of Mexico 2012 Expedition **2. Pinkspeckled shrimp**, NOAA/NMFS/SEFSC; **3. Barnacles** attached to a whale, Aleria Jensen, NOAA/NMFS/AKFS; **4. Squat lobster**, NOAA Okeanos Explorer Program, Gulf of Mexico 2012 Expedition; **5. Crayfish**, David Gerke, CC BY SA 3.0

Subphylum Chelicerata

Spiders, ticks, horseshoe crabs, and scorpions all belong to the subphylum Chelicerata. These arthropods have two body sections, most have four pairs of walking legs, and they lack the antennae found on most other arthropods. The mouthparts called chelicerae have fangs used to paralyze prey. A second pair of appendages called pedipalps are longer than the chelicerae and are used to grab the prey once it has been stabbed and paralyzed.

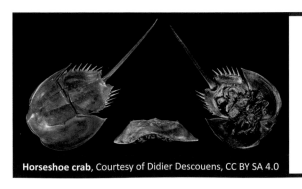

Horseshoe crab, Courtesy of Didier Descouens, CC BY SA 4.0

Class Merostomata

Horseshoe crabs have a hard external shell, chelicerae, and 5 pairs of legs. They live in marine environments and are not true crabs. Their anatomy seems to have changed very little from fossils dating back to 500 million years ago.

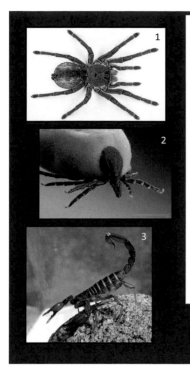

Class Arachnida

Order Araneae: Spiders

The largest group of arachnids; vary from pinhead to dinner plate size. Spiders stab, paralyze, and liquefy their prey by injecting digestive enzymes into it, and then they suck out the tissues of the prey.

Order Acari: Mites and ticks

These are parasitic arachnids that feed off a host's tissue and blood. Specialized chelicerae and pedipalps allow them to pierce the skin of their host and attach tightly to it.

Order Scorpiones: Scorpions

Scorpions live in warm areas. They have long enlarged pedipalps used for claws and a long segmented abdomen that carries a powerful stinger. Scorpions have jaws used to chew their prey.

Photo credits: 1. **Spider**–Goliath Birdeater *(Theraphosa blondi),* Didier Descouens, CC BY SA 4.0 ;
2. **Tick** *(Ixodus ricinus),* Richard Bartz, CC BY SA 2.5;
3. **Northern Scorpion** *(Paruroctonus boreus),* Xbuzzi, CC BY SA 4.0

Subphylum Uniramia

The subphylum Uniramia includes millipedes, centipedes, and all the insects. Uniramians are arthropods with jointed legs, unbranched appendages, and only one pair of antennae. Animals in this subphylum live in a wide variety of habitats and have body plans that are diverse. Centipedes and millipedes resemble worms with lots of legs, and insects come in a variety of shapes and sizes. These animals are found crawling, flying, or swimming everywhere in the world.

Photo credits:
1. **Millipede**, Thomas Shahan, CC BY SA 2.0;
2. **Millipede**, K. Schütte, CC BY SA 3.0
3. **Centipede**, Eran Finkle, CC BY SA 3.0

Class Diplopoda: Millipedes

Millipedes have two pairs of jointed legs on each of many body segments. Some will roll into a tight coil for protection when threatened. Millipedes eat decaying leaves and other dead plant material

Class Chilopoda: Centipedes

Centipedes have a single pair of jointed legs attached to each segment of the body. They are carnivores and use their chelicera and pedipalps to catch, poison, and kill their prey. They have rounded, flat heads, and many species lack eyes.

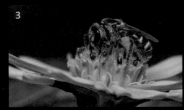

Photo credits: 1. **Leopard lacewing butterfly** *(Cethosia cyane)*, AirBete, CC BY SA 2.5;
2. **Carpenter ant** *(Camponotus species)* with aphids, Judy Gallagher, CC BY SA 2.0;
3. **Honeybee** *(Apis sp.)*, Yogendra Joshi, CC BY SA 2.0

Class Insecta

Insects are the largest group of animals, making up 73% of all animal life. Insects have a segmented body with jointed appendages, an exoskeleton, a body that is divided into three parts (head, thorax, abdomen), 3 pairs of legs attached to the thorax, a pair of antennae, two pairs of wings on the thorax, a pair of compound eyes, and tracheal tubes that carry air during respiration. Most undergo metamorphosis where their body structure transforms dramatically from a juvenile stage to a mature adult stage.

14.7 Phylum Echinodermata

Echinoderms include sea stars (starfish), sea urchins, sand dollars, sea cucumbers, and crinoids. The word echinoderm comes from two Latin words: *echino* which means "spiny," and *dermis* which means "skin." Echinoderms are marine animals with spiny skin.

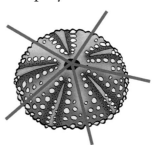

Echinoderms are unique among animals. Adult echinoderms generally have a five-part radially symmetrical body with no head or tail. The body is two-sided. The lower side is called the oral surface and contains the mouth and small tube feet used for movement.

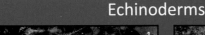
Echinoderms

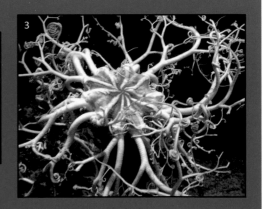

Photo credits: 1. Sea urchin, David Monniaux, CC BY SA 3.0;
2. Sea star, Derek Keats from Johannesburg, South Africa, CC BY SA 2.0;
3. Basket star, Smart Destinations, CC BY SA 2.0

The upper side is called the aboral surface and has a spiny exterior. (Aboral comes from the Latin words *ab,* meaning "away from," and *oris,* meaning "mouth.") Echinoderms also have internal hardened plates of calcium that act like an internal skeleton, called an endoskeleton.

14.8 Summary

● Sponges are in the phylum Porifera. They have asymmetrical bodies with no tissues or organ systems, and they pump water through a central cavity in order to filter out food.

● Jellyfish, hydras, sea anemones, and corals are in the phylum Cnidaria. Some distinguishing characteristics are radially symmetrical bodies and tentacles that have stinging cells used to capture prey.

● Worms are divided into three phyla: Platyhelminthes (flatworms), Nematoda (roundworms), and Annelida (earthworms, marine worms, and leeches).

● Snails, slugs, clams, squids, and octopuses are all in the phylum Mollusca, which is the second most abundant animal phylum on the planet. Mollusks are soft-bodied animals with an internal or external shell.

● The phylum Arthropoda includes crabs, lobsters, spiders, millipedes, insects, flies, and butterflies. Arthropods are animals with jointed legs and antennae and segmented bodies, and many have tough external skeletons made of chitin. Arthropods are the largest group of animals on the planet and are found everywhere.

● Echinodermata is the phylum for animals with spiny skin, including sea stars (starfish), sea urchins, sand dollars, sea cucumbers, and crinoids. An echinoderm has a two-sided body with the lower side containing the mouth and small tube feet and the upper side covered with spines. An echinoderm body has five-part radial symmetry.

14.9 Some Things to Think About

● Why do you think there are so many different kinds of non-chordates?

● Spend some time outdoors and look for worms, insects, and arachnids. Write your observations about these animals in your field notebook and make drawings of them.

● Why do you think echinoderms have radial symmetry?

Chapter 15 Chordates: Squirts, Amphibians, Reptiles, Birds

15.1 What Is a Chordate?

In the last chapter, we looked at non-chordate animals, which have a network of nerves throughout their bodies but don't have a vertebral column or a nerve cord. In this chapter we will look at chordates. Recall that chordates are animals that have a notochord during some stage of their development.

The word chordate comes from the Greek word *khorde* which means gut or string. A chordate is an animal that possesses *a string or cord (of nerves) along the back.* To be classified as a chordate, an animal must have a nerve cord on the back part of the body, a notochord (supporting rod running through the body below the nerve cord), pharyngeal pouches (tissues in the throat region that develop into gills in fishes and amphibians), and a tail that extends beyond the anus for at least some part of the animal's development. The classification of a chordate is based on the theory that all vertebrates have evolved from a common ancestor. The theory is that all chordates go through similar stages of embryonic development and therefore have similar features, such as a tail, pharyngeal pouches, and a notochord.

Glass catfish *(Kryptopterus vitreolus)*
In these transparent fish, the vertebral column and ribs are visible
Courtesy of Vassil

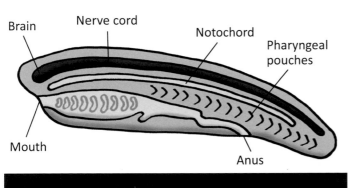

A simple chordate animal

Chordates come in a wide variety of shapes and sizes and have many different features. A chordate can look like a puffy pouch with holes in it; swim in the ocean with or without scales; have feet with toes, hooves, or webs; have fur or hair; or be completely bald. Some chordates live exclusively in water or on land, and others live in a habitat that is both watery and terrestrial. Some chordates reproduce by laying eggs, and some reproduce through live births. With so many variations, it's challenging to develop categories that fit the wide variety of animals in this phylum. In the next section, we'll take a closer look at the taxonomy of chordates.

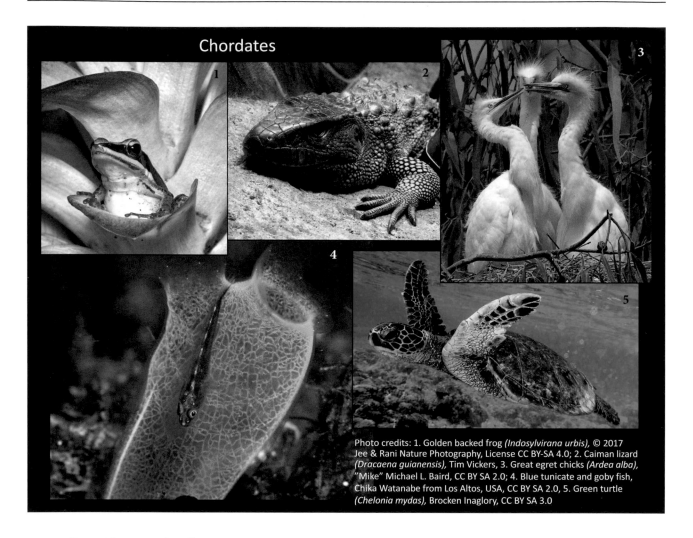

Chordates

Photo credits: 1. Golden backed frog *(Indosylvirana urbis),* © 2017 Jee & Rani Nature Photography, License CC BY-SA 4.0; 2. Caiman lizard *(Dracaena guianensis),* Tim Vickers, 3. Great egret chicks *(Ardea alba),* "Mike" Michael L. Baird, CC BY SA 2.0; 4. Blue tunicate and goby fish, Chika Watanabe from Los Altos, USA, CC BY SA 2.0, 5. Green turtle *(Chelonia mydas),* Brocken Inaglory, CC BY SA 3.0

15.2 Chordate Taxonomy

The phylum Chordata can be subdivided into three or four different subphyla. In this text, we will subdivide Chordata into the following four subphyla: Hemichordata, Urochordata, Cephalochordata, and Vertebrata. More than 99% of all chordates have a backbone (vertebral

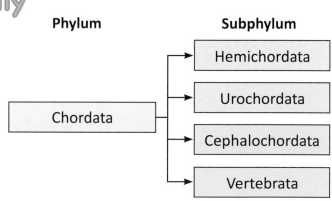

column) and are in the subphylum Vertebrata. The first three subphyla are non-vertebrates and include the 1% of chordates that lack a vertebral column (backbone) but do have a nerve cord that runs along the back part of the body. The subphylum Vertebrata contains all the animals with a vertebral column.

15.3 Subphylum Hemichordata

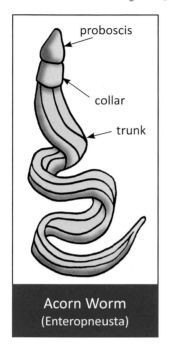

Acorn Worm
(Enteropneusta)

The subphylum Hemichordata is small and includes primarily the acorn worm and close relatives. There are around 130 known species in this subphylum. The prefix hemi- means "half"; hemichordates are animals that are halfway between full chordates and non-chordates.

Hemichordates live exclusively in marine environments. There are two main groups of hemichordates called Pterobranchia and Enteropneusta. The Pterobranchia are very small, up to 5 mm in length, and live in large colonies. Enteropneusts include acorn worms, and their sizes can range from less than a millimeter to more than two meters.

Hemichordates have three distinct body parts: a proboscis, a collar, and a long trunk. The notochord is found only in the proboscis and not in the remainder of the body, which is why they are called hemichordates and considered to be halfway between chordates and non-chordates.

15.4 Subphylum Urochordata (Tunicata)

The subphylum Urochordata, also called Tunicata, includes sea squirts, sea porks, salps, and sea tulips. There are just over 2000 species in this subphylum, living in the oceans and shallow waters. The prefix "uro-" comes from the Greek word *oura*, which means tail.

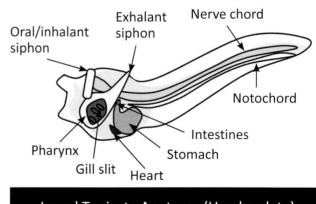

Larval Tunicate Anatomy (Urochordata)
Courtesy of Basketball171, CC BY SA 4.0

Tunicates begin as free-swimming tadpole-shaped larvae. The larva has all the characteristics of a chordate, with a notochord extending into the tail beyond the anus and pharyngeal tissues with gill slits. Adult tunicates lose their tail and notochord. Many attach to a solid surface for the remainder of their life, while some types swim freely in the ocean.

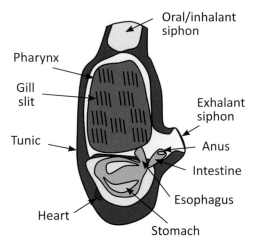

Oral/inhalant siphon

Pharynx

Gill slit

Tunic

Heart

Exhalant siphon

Anus

Intestine

Esophagus

Stomach

Adult tunicates have a simple body structure and specialized tissues and organs, including a heart, stomach, intestine, and reproductive organs. The barrel-shaped exterior is called the tunic and contains a large pharynx, which is the cavity behind the mouth opening where oxygen and food enter. Food and water are filtered through the pharynx, which is used for feeding and gas exchange, and wastes are expelled through the anus and siphon.

↑ Adult Tunicate Anatomy (above)
Courtesy of Basketball171, CC BY SA 4.0

Photo credits: 1. Sea pork (*Amaroucium stellatum*), Andrea Westmoreland from DeLand, United States, CC BY SA 2.0; 2. Sea squirt (*Polycarpa aurata*), Nick Hobgood, CC BY SA 3.0; 3. Sea tulip (*Pyura spinifera*), Richard Ling from NSW, Australia, CC BY SA 2.0

Tunicates

15.5 Subphylum Cephalochordata

Lancelet
Courtesy of © Hans Hillewaert, CC BY SA 4.0

The subphylum Cephalochordata includes the lancelets, which are small fish-like animals that live on the sandy ocean floor. The prefix cephalo- comes from the Greek word *kephale* which means "head." Animals in this subphylum have a notochord that extends into the head.

Lancelets have specialized tissues and organs, including a mouth, intestines, muscles, and reproductive organs. However, lancelets have no brain or other head parts. Unlike tunicates, the pharynx in a lancelet is used exclusively for filter feeding and not for gas exchange. Also, lancelets lack a heart but do have a closed circulatory system.

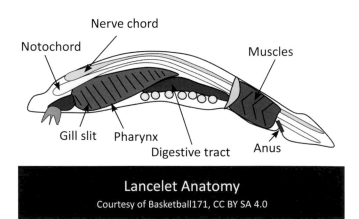

Nerve chord

Notochord

Muscles

Gill slit   Pharynx

Digestive tract   Anus

Lancelet Anatomy
Courtesy of Basketball171, CC BY SA 4.0

15.6 Subphylum Vertebrata

The subphylum Vertebrata has over 50,000 species of animals and includes lampreys, fish, sharks, amphibians, reptiles, birds, and mammals. Because there are so many different kinds of vertebrates with a wide variety of different features, the subphylum Vertebrata is further divided into different classes. Scientists disagree on how many classes are needed to divide Vertebrata. Different references cite anywhere from four to nine separate classes. In this text we will use seven classes for grouping the vertebrates.

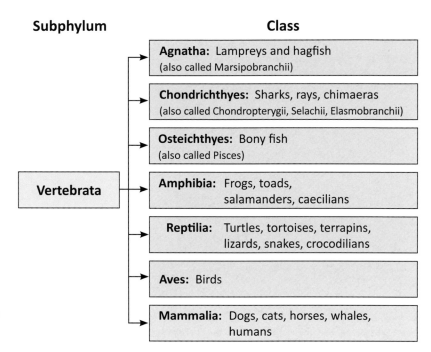

Subphylum	Class
	Agnatha: Lampreys and hagfish (also called Marsipobranchii)
	Chondrichthyes: Sharks, rays, chimaeras (also called Chondropterygii, Selachii, Elasmobranchii)
	Osteichthyes: Bony fish (also called Pisces)
Vertebrata	**Amphibia:** Frogs, toads, salamanders, caecilians
	Reptilia: Turtles, tortoises, terrapins, lizards, snakes, crocodilians
	Aves: Birds
	Mammalia: Dogs, cats, horses, whales, humans

We will cover the first six classes of the subphylum Vertebrata in this chapter and discuss the last class, Mammalia, in the next chapter.

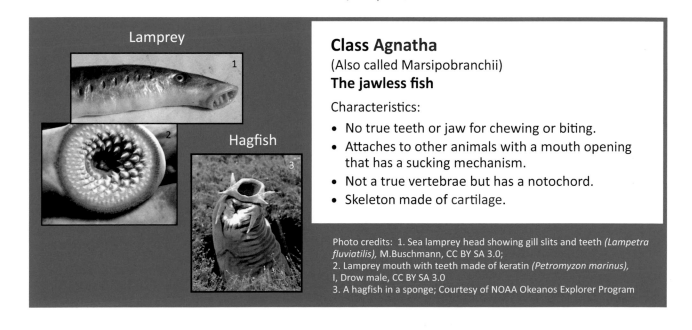

Lamprey

Hagfish

Class Agnatha
(Also called Marsipobranchii)
The jawless fish

Characteristics:

- No true teeth or jaw for chewing or biting.
- Attaches to other animals with a mouth opening that has a sucking mechanism.
- Not a true vertebrae but has a notochord.
- Skeleton made of cartilage.

Photo credits: 1. Sea lamprey head showing gill slits and teeth *(Lampetra fluviatilis)*, M.Buschmann, CC BY SA 3.0;
2. Lamprey mouth with teeth made of keratin *(Petromyzon marinus)*, I, Drow male, CC BY SA 3.0
3. A hagfish in a sponge; Courtesy of NOAA Okeanos Explorer Program

Class Chondrichthyes
(also called Selachii, Chondropterygii, Elasmobranchii)
Sharks, Rays, Skates, Sawfish

Animals in this class have flexible cartilage skeletons, specialized tissues and organs, and gills on both sides of the body. Some have live births and others lay soft-shelled eggs.

1. **Sharks:** Best known for their many sharp teeth, which are constantly being lost and replaced. Have distinctive fins.

2. **Rays and Skates:** Have a flattened body, 5-6 gill openings and a mouth on the underside of the body, and long slender whiplike tails or thick tails.

3. **Sawfish:** A type of ray with a saw-like elongated rostrum edged with teeth. The nostrils, mouth, and gills are located on the underside of the body.

Photo credits: 1. Great White Shark, Elias Levy, CC BY SA 2.0; 2. Sawfish, J. Patrick Fischer, CC BY SA 3.0; 3. Ray, Public Domain

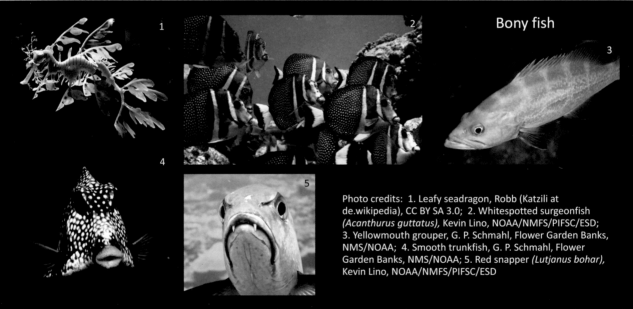

Bony fish

Photo credits: 1. Leafy seadragon, Robb (Katzili at de.wikipedia), CC BY SA 3.0; 2. Whitespotted surgeonfish *(Acanthurus guttatus)*, Kevin Lino, NOAA/NMFS/PIFSC/ESD; 3. Yellowmouth grouper, G. P. Schmahl, Flower Garden Banks, NMS/NOAA; 4. Smooth trunkfish, G. P. Schmahl, Flower Garden Banks, NMS/NOAA; 5. Red snapper *(Lutjanus bohar)*, Kevin Lino, NOAA/NMFS/PIFSC/ESD

Class Osteichthyes (also called Pisces):The Bony fish

With over 29,000 species, Osteichthyes is the largest class of vertebrates. They are found in both fresh and saltwater environments and have hard bone skeletons made of calcium. Most are in a group called ray-finned fish that have slender bony spines connected to a thin layer of skin that forms their fins. Bony fish have specialized tissues and organs, a number of specialized fins, gills for oxygen exchange, and scales covered in a slimy mucus that helps protect them from predators. A swim bladder that they can fill and deflate allows them to swim at a variety of water depths.

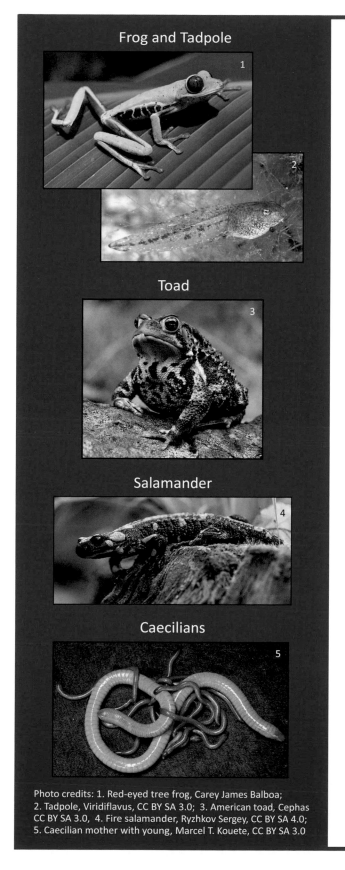

Frog and Tadpole

Toad

Salamander

Caecilians

Photo credits: 1. Red-eyed tree frog, Carey James Balboa; 2. Tadpole, Viridiflavus, CC BY SA 3.0; 3. American toad, Cephas CC BY SA 3.0, 4. Fire salamander, Ryzhkov Sergey, CC BY SA 4.0; 5. Caecilian mother with young, Marcel T. Kouete, CC BY SA 3.0

Class Amphibia
Frogs and Toads, Salamanders, Caecilians

Many amphibians begin life in the water as free-swimming tadpoles, and as adults they live both on land and in the water. Amphibia is a relatively small class with just under 8000 species. Amphibians come in many different sizes and colors and have specialized tissues and organs. Juveniles breathe using gills, which are replaced by lungs in many adult amphibians. Some adults never develop lungs and breathe exclusively through their skin and mouth. Others live in moist environments and keep their gills through adulthood.

Order Anura: Frogs and Toads

With over 6000 species, frogs and toads are the most abundant group of amphibians. Adults have a short body and four legs—two smaller legs in front and two larger legs in back. Frogs and toads are similar, but frogs live nearer to water environments. Toads can live in drier environments, such as moist woods or even a dry desert. Frogs are often more brightly colored than toads. Both lay their eggs in water and begin life as a tadpole.

Order Caudata: Salamanders

Adults have a long narrow body, a blunt snout, and a tail. Most have two or four short legs, and they come in different colors and sizes. Salamanders are carnivores and eat other animals, such as worms, slugs, snails, and small fish. They lay eggs in water environments, and their young resemble tadpoles.

Order Gymnophiona: Caecilians

A small group of about 200 known species. Caecilians have a long body and no legs, making them look like worms or snakes. They live in a water or moist environment, do not have scales, and have no tail. They do have a backbone and skull.

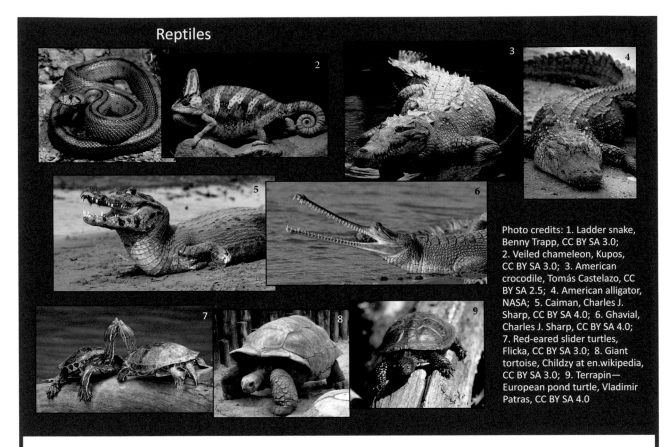

Reptiles

Photo credits: 1. Ladder snake, Benny Trapp, CC BY SA 3.0; 2. Veiled chameleon, Kupos, CC BY SA 3.0; 3. American crocodile, Tomás Castelazo, CC BY SA 2.5; 4. American alligator, NASA; 5. Caiman, Charles J. Sharp, CC BY SA 4.0; 6. Ghavial, Charles J. Sharp, CC BY SA 4.0; 7. Red-eared slider turtles, Flicka, CC BY SA 3.0; 8. Giant tortoise, Childzy at en.wikipedia, CC BY SA 3.0; 9. Terrapin— European pond turtle, Vladimir Patras, CC BY SA 4.0

Class Reptilia: Snakes, Lizards, Crocodiles, Turtles, Tortoises

There are over 10,000 reptile species. They have a vertebral column, dry scaly skin, and lungs. They lay eggs on land and can live entirely on land without the need for an aquatic environment.

Reptiles have specialized tissues and organs and eat a variety of foods. Some are herbivores and others are carnivores. Reptiles are ectotherms and regulate their body temperature by using their environment to warm or cool their bodies, seeking out sun or shade.

Order Squamata—Lizards and Snakes

Lizards have legs, movable eyelids, clawed toes, and external ears. Snakes are legless, moving along the ground or through water by using their muscles and the scales on their skin to "walk" or swim.

Order Crocodilia—Alligators, Crocodiles, Caimans, Gavials

In general, crocodilians have a long snout and numerous jagged teeth, are carnivorous, lay eggs, nurture their young, and live in warm tropical or subtropical watery environments. Alligators generally live in freshwater environments and crocodiles in saltwater marshes and swamps. Caimans live in Central and South America. Gavials have very long, slender jaws with sharp teeth used to catch fish. They live in the rivers of India and Nepal.

Order Testudines (also called Chelonia)—Turtles, Tortoises, and Terrapins

These animals have a hard, dome-shaped external shell. Turtles live in water most of the time, have webbed feet, and some, like the sea turtle, leave the water only to lay eggs in the sand. Tortoises are land animals with short stumpy legs for walking, and terrapins live both in the water and on land.

Birds

Class Aves: Birds

All birds have wings, two legs covered in scales, and an outer covering of feathers. Birds are able to keep a constant internal temperature, generating their own body heat. Animals that can generate their own body heat are called endotherms.

Birds eat a variety of foods. Some are carnivorous and eat primarily animals; others are herbivores and live on plants and seeds. Because birds don't have teeth, they cannot chew food when they eat. To help aid digestion, many birds have a specialized organ called a crop that is an enlarged part of the esophagus, the tube that connects the mouth and stomach. Food is stored in the crop and moistened for digestion. Birds that eat insects and seeds have an additional organ called a gizzard that helps break down food by grinding. Some birds swallow small bits of gravel that are used in the gizzard to help with the grinding of food. The gizzard is a small organ with powerful muscles and sits just above the small intestine.

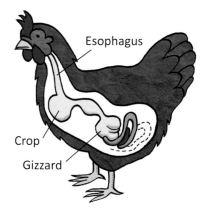

There are over 30 different orders of birds. Birds vary greatly in size and shape, and although most birds fly, some fly very little or not at all. The ostrich is a type of flightless bird that moves by running or sometimes swimming, and the penguin is a flightless bird that is adapted to living in a cold water environment.

15.7 Summary

- A chordate is an animal that has a nerve cord along the back part of the body, a notochord, pharyngeal pouches, and a tail that extends beyond the anus for at least some part of the animal's development.

- Animals with a backbone are called vertebrates and are in the subphylum Vertebrata.

- The subphylum Hemichordata includes the acorn worm and close relatives, which are considered to be halfway between full chordates and non-chordates because the notochord is found only in the proboscis and not in the remainder of the body.

- The subphylum Urochordata, also called Tunicata, includes sea squirts, sea porks, salps, and sea tulips. Adult tunicates have a simple body structure and specialized tissues and organs.

- The subphylum Cephalochordata includes the lancelets, which are small fish-like animals that have a notochord that extends into the head but no vertebral column.

- The subphylum Vertebrata has over 50,000 species of animals and includes lampreys, bony fish, sharks, amphibians, reptiles, birds, and mammals. Vertebrata is further divided into several classes. All the animals in this subphylum have a backbone.

15.8 Some Things to Think About

- How are the non-vertebrate chordates similar? How are they different?

- Why do you think sharks have so many teeth?

- Why do you think some fish are brightly colored and other fish are brown or silver?

- Why do you think amphibians start out as tadpoles?

Chapter 16 Chordates: Mammals

Photo credits:
See end of chapter

16.1 What Is a Mammal?

From the tiny Etruscan shrew to the gigantic blue whale, animals in the class Mammalia inhabit both land and sea. Although there are over 5000 different species of mammals of various sizes, shapes, and colors, mammals share a few common features. All mammals have fur at some point in their life, breathe air through lungs, nurse their young with mammary glands, and are endothermic, regulating their own internal body heat.

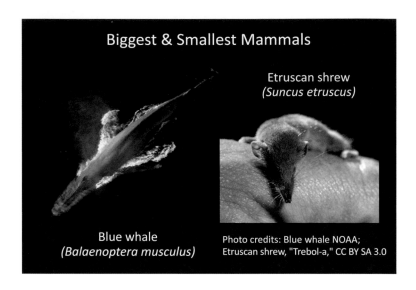

Biggest & Smallest Mammals

Etruscan shrew
(Suncus etruscus)

Blue whale
(Balaenoptera musculus)

Photo credits: Blue whale NOAA;
Etruscan shrew, "Trebol-a," CC BY SA 3.0

Mammals also have specialized tissues and organs, and because they have adapted to where they live and what they eat, mammals have features that are unique to their species. For example, the jaws and teeth of herbivores, or plant eating mammals, differ from the jaws and teeth of carnivores, or meat-eating mammals. Herbivores such as horses, cows, and deer have chisel-like front teeth called incisors for tearing plants, and broad flattened molars in back for grinding. On the other hand, carnivores, such as wolves, lions, and tigers, have pointed canine teeth used for piercing and tearing flesh and ridged sharp molars for chewing and slicing through meat. Many herbivores, such as cattle, have a stomach with four compartments, including a rumen where cellulose is digested.

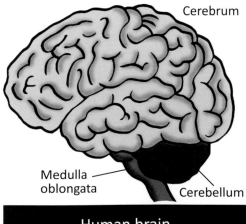

Cerebrum

Medulla
oblongata

Cerebellum

Human brain

Mammals also have the most developed brain of any of the animals. In general, mammals have large brains compared to their body size. Mammalian brains are composed of three main sections: the cerebrum, cerebellum, and medulla oblongata.

The cerebrum allows mammals to process complex thinking and learning, while the cerebellum controls muscular coordination. Breathing, heart rate, and other involuntary body functions are controlled by the medulla oblongata.

16.2 Taxonomy

Given the wide diversity of mammals and their varied and often specialized features, figuring out how to organize them into taxonomic groups can be challenging. It is important to understand that the field of taxonomy is a dynamic science and changes constantly. Modern classification schemes generally try to reflect ancestral relationships between organisms, and although modern technologies like DNA sequencing can help determine possible ancestral connections, it has also made the science of taxonomy increasingly complicated.

Early mammalian classification and taxonomy were created by the Swedish botanist and physician Carolus Linnaeus in the mid 1700s. Many of Linnaeus's classifications are still used today, and although there have been some significant revisions to Linnaeus' classifications, no one system of classification is universally accepted. In 1997 the classification of mammals was revised by Malcolm McKenna and Susan Bell, which resulted in the McKenna/Bell classification. In 2005 additional revisions were introduced by Don H. Wilson and DeeAnn M. Reeder that expanded the number of mammalian subspecies, resulting in the Wilson/Reeder classification system.

With the revised taxonomy in mind and to keep it simple, we will use three main divisions of mammals: monotremes, marsupials, and placentals. Monotremes are mammals that lay eggs, marsupials are generally characterized by the female having a pouch in which it rears its young through early infancy, and placentals generally nourish their young during gestation via an organ called a placenta. We will take quick look at a few species of monotremes and marsupials and explore a larger variety of placental species.

16.3 Order Monotremata

Monotremes lay soft-shelled eggs that incubate and hatch outside the body, similar to reptiles. However, unlike reptiles, young monotremes are nourished with mother's milk coming from mammary glands. There are only three genera of monotremes: the platypus and two genera of echidnas. These animals are found in Australia and New Guinea.

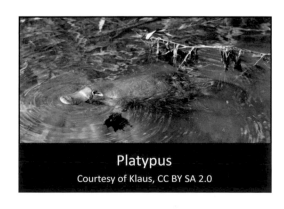

Platypus
Courtesy of Klaus, CC BY SA 2.0

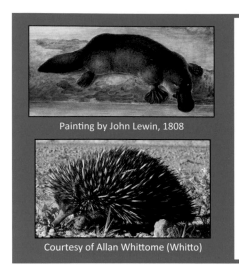

Painting by John Lewin, 1808

Courtesy of Allan Whittome (Whitto)

Genus *Platypus*

The platypus has an otter-like body with fur, a beaver-like tail, and a flat, often bluish-colored bill. The male has sharp, venomous spurs behind its webbed feet. The platypus is carnivorous, eating freshwater shrimp, insects, and worms.

Genus *Echidna*

Also called spiny anteaters, echnidas are egg-laying mammals covered with fur and protective spines. There are two genera of echidnas: the long-beaked echidna and the short-beaked echidna. Echidnas have short legs and feet with claws used for digging to find insects and grubs, and they have a very small mouth with a long, thin, sticky tongue used to catch food.

16.4 Order (Infraclass) Marsupialia

Marsupials are an order (or infraclass[1]) of mammals that rear their young in a pouch. Kangaroos, opossums, wombats, and koalas are examples of marsupials. A young marsupial begins life as an embryo in the mother's reproductive tract, growing and feeding from a small yolk sack. Once the food in the yolk sac has been consumed, the baby marsupial crawls to a fur-lined pouch called the marsupium and drinks milk from the mother's nipple. It can spend several months in the pouch until it is old enough to survive on its own.

Koala
(Phascolarctos cinereus)
Courtesy of Quartl, CC BY SA 3.0

Kangaroo with joey (baby) in pouch
Courtesy of Graeme Rainsbury

16.5 Order (Infraclass) Eutheria/Placentalia

Placental mammals are those mammals that bear live young. The offspring of these mammals have been nourished inside a special organ called the placenta that forms when tissues from the embryo join with the tissues of the mother. A placenta allows the exchange of nutrients, oxygen, and carbon dioxide between the mother's body and the embryo, creating a space for the embryo to develop over long periods of time. The gestation time, or

1. Marsupialia and Placentalia are now generally referred to as an infraclass; however, some resources continue to list them as orders.

time the offspring stays inside the placenta, varies between species. Baby mice develop for only a few weeks before being born, while elephants take up to two years! Placental animals include cats, dogs, whales, humans, and sheep.

Given the huge diversity of animals, biologists are still trying to figure out how to divide them into orders. Some resources cite nine different orders and other resources cite as many as 20. In this text we will take a look at 11 main orders of placental mammals and one superorder.

Order Artiodactyla

Artiodactyls are the even-toed ungulates and include cattle, goats, sheep, camels, pigs, hippopotamuses, and giraffes. The word ungulate means "hoofed animal." These mammals have hooves with an even number of digits on each foot. Two main types of hooves are found in Artiodactyla: a hoof with two weight-bearing toes

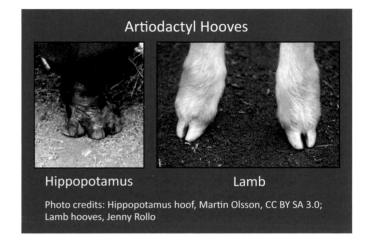

Artiodactyl Hooves

Hippopotamus   Lamb

Photo credits: Hippopotamus hoof, Martin Olsson, CC BY SA 3.0; Lamb hooves, Jenny Rollo

(sometimes called a cloven hoof) and a hoof with four toes that are spread out. There are over 200 species of artiodactyls in a wide range of sizes, and most mammals in this group are herbivores, eating primarily plants.

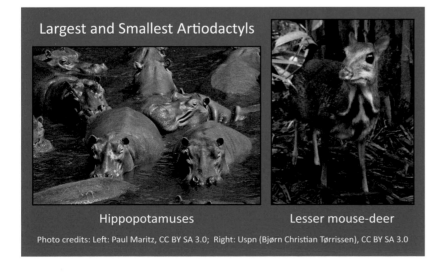

Largest and Smallest Artiodactyls

Hippopotamuses   Lesser mouse-deer

Photo credits: Left: Paul Maritz, CC BY SA 3.0; Right: Uspn (Bjørn Christian Tørrissen), CC BY SA 3.0

Artiodactyls are widespread and found on every continent except Antarctica (although it should be noted that artiodactyls were introduced to Australia and New Zealand by humans). Artiodactyls live in a variety of habitats including forests, deserts, grasslands, savannas, tundras, and mountains.

Order Perissodactyla

Perissodactyls are the odd-toed ungulates and include horses, zebras, rhinoceroses, and tapirs. These mammals have an odd number of digits, or toes, on each foot. They have long, tapered faces, large nostrils, and are exclusively land animals.

Mammals in this group are herbivores and eat only plants. They have a simple stomach, and unlike even-toed hoofed mammals, they do not have a rumen. Instead food travels from the stomach through the small intestine to a pouch called a cecum where it is broken down by bacteria before going into the large intestine.

Perissodactyl Hooves

1. Underside of hooves of sleeping tapir – Left: 4-toed front hoof, Right: 3-toed back hoof; Tapirs are classified by the structure of their back hooves; 2. Horse's one-toed hoof; 3. Underside of horse's hoof; 4. Three-toed front hoof of sleeping rhinoceros

Photo credits: 1. Sasha Kopf, CC BY SA 3.0; 2. Montanabw, CC BY SA 3.0; 3. Alex Brollo, CC BY SA 4.0; 4. Jose Luis Navarro, Public Domain

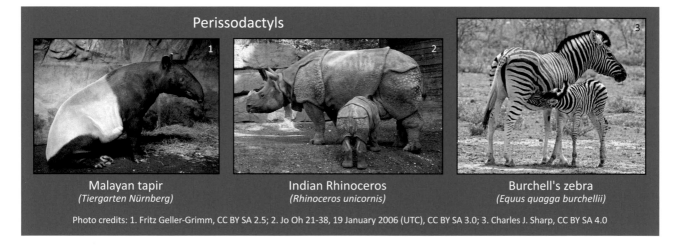

Perissodactyls

Malayan tapir
(Tiergarten Nürnberg)

Indian Rhinoceros
(Rhinoceros unicornis)

Burchell's zebra
(Equus quagga burchellii)

Photo credits: 1. Fritz Geller-Grimm, CC BY SA 2.5; 2. Jo Oh 21-38, 19 January 2006 (UTC), CC BY SA 3.0; 3. Charles J. Sharp, CC BY SA 4.0

Order Carnivora

Carnivores get their food primarily by eating other animals. Animals in this order include cats, tigers, dogs, wolves, bears, foxes, raccoons, and walruses. Some are predators and eat living animals, while others are scavengers and eat dead animals. There are a few in this group, such as the brown bear, that are omnivores and eat both meat and plants. The Giant Panda is grouped with the carnivores but eats primarily bamboo and other plants and very little meat. The order Carnivora is separated into two suborders: Feliformia (cats, hyenas, mongooses, meerkats, civets, genets) and Caniformia (dogs, wolves, bears, mustelids, seals). There are over 270 species of carnivores of various shapes and sizes. Most animals in this group are terrestrial and live on land, with a few species living in the oceans.

Feliformia Carnivores

Photo credits: 1. Caracal *(Caracal caracal)*, Leo za1, CC BY SA 3.0; 2. Binturong *(Arctictis binturong)*, Tim Strater from Rotterdam, Nederland, CC BY SA 2.0; 3. Meerkats *(Suricata suricatta)*, Sara&Joachim&Mebe, CC BY SA 2.0

Caniformia Carnivores

Photo credits: 1. Sea otter *(Enhydra lutris)*, Marshal Hedin from San Diego, CC BY SA 2.0; 2. Raccoon *(Procyon lotor)*, Darkone, CC BY SA 2.5; 3. Kodiak bear *(Ursus arctos middendorffi)*, Yathin S. Krishnappa, CC BY SA 3.0; 4. Mexican wolf *(Canis lupus baileyi)*, Jim Clark, USFWS

Order Cetacea

Cetaceans include about 90 species of whales, dolphins, and porpoises. Most of these are marine mammals, spending all their time in the oceans, but a few freshwater species live in rivers. Like land mammals, cetaceans nurse their young and use lungs to breath air, requiring them to surface frequently. Air is taken in through one or two blowholes on the top of the head. Cetaceans give birth to live young, have traces of hair or fur, are carnivorous, and are endothermic, with a layer of fat, or blubber, under the skin to help keep them warm.

Cetaceans

Photo credits: 1. Bottlenose dolphin, NASA; 2. Humpback whale with her calf, National Marine Sanctuaries, CC BY SA 2.0

Order Lagomorpha

Many people are familiar with rabbits and their long ears, soft fur, and powerful hind legs used for leaping. Lagomorphs are herbivores and eat only plants. Lagomorpha is divided into two basic groups: Leporidae (hares and rabbits) and Ochotonidae (pikas).

Hares and rabbits are small animals with short tails, long ears, and long hind legs. They have fur on the soles of their feet, which gives them added traction when running. Hares and rabbits have acute hearing and good night vision, both adaptations to the crepuscular (twilight) and nocturnal (nighttime) lifestyles of many of the species in this group.

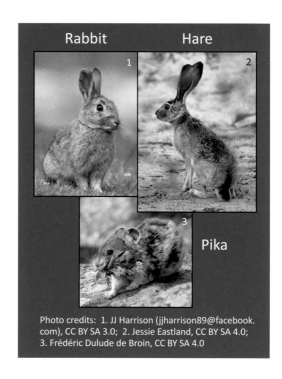

Rabbit   Hare

Pika

Photo credits: 1. JJ Harrison (jjharrison89@facebook.com), CC BY SA 3.0; 2. Jessie Eastland, CC BY SA 4.0; 3. Frédéric Dulude de Broin, CC BY SA 4.0

Elephants

Elephants communicating—Courtesy of jinterwas, CC BY SA 2.0

Order Proboscidea

The proboscideans are animals with trunks—the elephants. A proboscis is an arm "in front" that is used for "feeding." Proboscideans use a "front arm," or trunk, to pick up food, draw up water, and even move large objects.

Order Rodentia

The word Rodentia comes from the Latin verb *rodere,* which means "to gnaw." Animals in this order include squirrels, mice and rats, and they all are chewers. They gnaw everything from wood to paper to your favorite pair of shoes, if they can! These animals have a continuously growing pair of incisors (front teeth) on the upper and lower jaws that require gnawing to wear them down and keep them from getting too long. Most rodents are herbivores, but some are omnivores, while others are insect-eating carnivores.

Rodentia is the largest group of mammals and includes about 1500 species with a variety of characteristics. They are generally small in size with short limbs and long tails. Some live in burrows in the ground, while others make their homes in trees or spend part of their time in water.

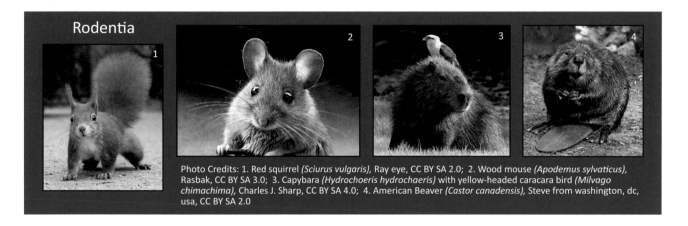

Rodentia

Photo Credits: 1. Red squirrel *(Sciurus vulgaris),* Ray eye, CC BY SA 2.0; 2. Wood mouse *(Apodemus sylvaticus),* Rasbak, CC BY SA 3.0; 3. Capybara *(Hydrochoeris hydrochaeris)* with yellow-headed caracara bird *(Milvago chimachima),* Charles J. Sharp, CC BY SA 4.0; 4. American Beaver *(Castor canadensis),* Steve from washington, dc, usa, CC BY SA 2.0

Order Eulipotyphla (Insectivora)

The order Eulipotyphla is a redefined order of the now abandoned order Insectivora. The order Eulipotyphla includes hedgehogs, moonrats, certain shrews, moles, and shrew moles. Animals in this order tend to be small, insect-eating animals. They often live underground and have an excellent sense of smell and hearing.

Hedgehogs are spiny, resembling the porcupine, which is a rodent. Unlike porcupines, the spines of a hedgehog do not detach from the body and are not barbed. However, both hedgehogs and porcupines roll into a ball with the spines pointing outward as a defense.

Shrews are very small animals that tend to be fierce and that produce substances that taste bad to help protect them from predators. Moles live underground, have very poor eyesight or are blind, and have strong, clawed feet for digging.

Eulipotyphla

Photo Credits: 1. Mole *(Scalopus Aquaticus)*, Kenneth Catania, Vanderbilt University, CC BY SA 3.0; 2. Hedgehog *(Erinaceus europaeus)*, © Michael Gäbler/Wikimedia Commons/CC BY-SA 3.0; 3. Young hedgehog—rolled up, Jacek Zapała (Jacek (talk)), CC BY SA 4.0; 4. Common shrew *(Sorex araneus)* eating an earthworm, Soricida, CC BY SA 3.0

Superorder Xenarthra

The name Xenarthra comes from the Greek word roots *xenos,* which means "foreign, or alien," and *arthron,* which means "joint." Xenarthrans are mammals that have "strange or foreign joints" in their lower spine. The superorder Xenarthra used to be called Edentata, which means "without teeth." The mammals in this superorder don't have front teeth, but most have very small teeth found at the back of the jaw. Anteaters are an exception because they don't have any teeth.

Xenarthra

Photo Credits: 1. Two-toed sloth *(Choloepus hoffmanni)*, Geoff Gallice, CC BY SA 2.0; 2. Giant anteater *(Myrmecophaga tridactyla)*, Dave Pape, Public Domain; 3. Nine-banded armadillo *(Dasypus novemcinctus)*, Vlad Lazarenko, CC BY SA 3.0

Order Chiroptera

Chiroptera is the order for the only winged mammals capable of flight, the bats. The word Chiroptera comes from two Greek words, *cheir* or "hand" and *pteron* "wing." Bats have true powers of flight, unlike flying squirrels and other mammals that can jump and glide but do not possess the mechanics for actual flight. Bats have been adapted for flight with long,

Bats

Photo credits: 1. Flying fox bat sleeping, Nobbi, CC BY SA 3.0; 2. Big eared townsend bat, National Park Service; 3. Common Fruit Bat, Original photo—Oren Peles, Derivative work—User-MathKnight, CC BY SA 2.5

spindly arms, extremely long fingers, and membranes that stretch from the bat's body to the ends of the fingers and between them.

Manatee

Dugong

Photo credits: Manatee—Jim P. Reid, USFWS; Dugong—Camille Ménard, CC BY SA 3.0

Order Sirenia

Mammals in the order Sirenia are large and slow-moving and sometimes called sea cows. There are four species in the order Sirenia: the West Indian manatee, the Amazonian manatee, the West African manatee, and the dugong. These mammals are herbivores found in shallow, slow-moving rivers, bays, and warm coastal swamps, and they can move freely between fresh and salt water.

Manatees and dugongs are similar and different. Manatees have large gray bodies with a flat paddle-shaped tail. Dugongs have a similar body shape and color but have tail flukes with pointed projections, more closely resembling the tail of a whale. All of these animals need to surface to breathe air, and they close their nostrils when submerged to keep water out. Some can stay under water for as long as 20 minutes, but they usually surface every 2 to 5 minutes to breathe.

Order Primates

The order Primates has over 300 species and includes monkeys, lemurs, apes, tarsiers, and humans. The word primate comes from the Latin word *primas,* meaning "prime, first rank." The order Primates is divided into two suborders: Strepsirrhini (the "wet-nosed" animals–lemurs, galagos, and lorisids) and Haplorhini (the "dry-nosed" animals–tarsiers, monkeys, apes, and humans).

Primates vary in size from the tiny Madame Berthe's mouse lemur at 30 g (1 oz.) to the huge eastern lowland (or Grauer's) gorilla that can weigh over 200 kg (440 lb.). Most primates spend all or part of their lives in trees, and all have tails, except for apes and humans. Primates tend to have larger brains relative to body weight than other land animals. Most have flat nails rather than hooves or claws, and even those with claws

Primates

Photo credits: 1. Ring-tailed lemur, Mathias Appel, Public Domain; 2. Tarsier, mtoz, CC BY SA 2.0; 3. Baboons, Till Niermann, CC BY CA 3.0; 4. Macaques, Yblieb, CC BY SA 3.0; 5. Golden-headed Lion Tamarin, Hans Hillewaert, CC BY SA 3.0

have at least one flat nail. Some primates have hands and feet. Most primates live in tropical or subtropical environments. Humans are the only primate that lives on all continents.

Humans, apes, and a few other primates have opposable thumbs that allow them to grasp and manipulate objects. An opposable thumb is one that can reach around to touch the other fingers on the same hand. Primates also have ridges of skin on the hands and feet that can leave fingerprints.

All primates have two eyes that face forward, providing overlapping visual fields for stereoscopic vision. In this type of vision the two eyes view the same object from slightly different angles, and the brain puts the two images together into one. This gives the animal the ability to accurately gauge distances.

16.6 Human Anatomy

Human anatomy is the study of the specific body parts that make up the human body. There are two ways to study the human body: gross anatomy and microscopic anatomy. Gross anatomy is the study of systems and organs visible to the eye. Microscopic anatomy involves using a microscope to examine very small features. Biologists use both gross anatomy and microscopic anatomy to get a full picture of how the human body works.

Like all animals, the human body is made up of cells, tissues, and organs that are organized into different organ systems. There are 11 different organ systems that make up the human body, as follows:

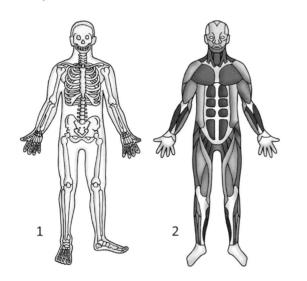

1. Skeletal system: Consists of bones and connective tissue; supports the body and protects internal organs.

2. Muscular system: Made up of groups of muscles attached to the bones of the skeleton; movement is produced as muscles contract or relax.

3. Circulatory system: The heart and the network of vessels that transport blood throughout the body, carrying oxygen and nutrients to cells and removing waste products from cells.

4. Nervous system: The main control system of the body; the brain, the spinal cord, and the network of nerves that go through the body.

5. Lymphatic (immune) system: Made up of a network of vessels; protects the body from infections by collecting fluid from the tissues and delivering it to the blood.

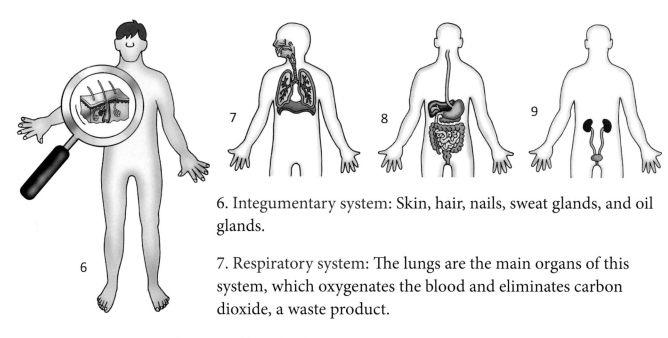

6. Integumentary system: Skin, hair, nails, sweat glands, and oil glands.

7. Respiratory system: The lungs are the main organs of this system, which oxygenates the blood and eliminates carbon dioxide, a waste product.

8. Digestive system: Takes in and breaks down foods for the body to use as fuel.

9. Excretory system: Used by the body to eliminate waste products from the cells through urine and feces.

10. Endocrine system: Made up of glands that release hormones that control body functions such as growth and energy production.

11. Reproductive system: Produces sperm (male) and eggs (female) used to create new organisms.

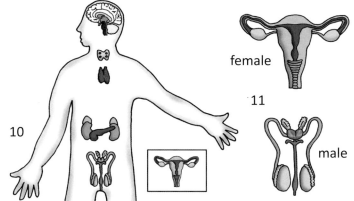

Having so many different systems allows an organism to attain homeostasis. Homeostasis is the state in which an organism maintains a controlled, stable internal environment. For example, our cells require a certain narrow temperature range. If the body is too hot or cold, our cells cannot function properly, which can cause permanent injury or even death. Our body maintains its optimal narrow temperature range by having different systems that heat our body or allow it to cool. For example, if you are unlucky enough to catch a virus, your body temperature will increase to kill the virus. However, your body will also sweat in order to cool your body down and keep it from overheating. Also, if you stay outside too long on a cold winter day, your body temperature could drop. Your body would respond by shivering, which allows your muscles to release heat, bringing your body temperature back up.

16.6 Summary

- All mammals have fur at some point in their life, breathe air through lungs, nurse their young with mammary glands, and are endothermic, regulating their own internal body heat.

- There are three major subdivisions of mammals: monotremes, marsupials, and placentals.

- There are different classification systems for placental animals. In this text 11 main orders are used for classification.

- Human anatomy is the study of the specific systems that make up the human body.

- Gross anatomy refers to the study of features of an organism that are large enough that a microscope is not required.

- Microscopic anatomy refers to the study of features that are too small to be seen without a microscope.

16.7 Some Things to Think About

- What advantages and disadvantages do mammals have over non-mammals.

- Why do you think it is difficult to classify mammals?

- How do you think DNA analysis might help or further complicate the classification of mammals?

- If you were to study anatomy, would you like to study gross anatomy or microscopic anatomy? Or a bit of both?

190   *Focus On Middle School Biology* 3rd Edition

Glossary-Index

[Pronunciation Key at end]

aboral surface (a-'bôr-əl 'sər-fəs) • a surface that is situated opposite to or away from the mouth, 163

absorb (əb-'sôrb) • in physics, for a light wave to have its energy transferred to a material, 97-99, 100-101

Acari (ak-'a-rē) • the order for mites and ticks in the class Arachnida, 161

acoelomate (ā-'sē-lə-māt) • a worm that does not have a coelom, 154

actin ('ak-tən) • a protein that forms long flexible filaments, 141, 143

adenosine triphosphate (ATP) (ə-'de-nə-sēn trī-'fäs-fāt) • an important energy molecule, 35-36, 46, 100-102, 141

AFM • see microscope, atomic force

agar ('ä-gər) • a gelatinous substance made of algae that is used in growing and testing bacteria and other organisms, 17

agar ('ä-gər) **plate** • a dish containing agar on which microorganisms are grown in a lab, 17

agar solution ('ä-gər sə-'lü-shən) • a mixture made with agar; used in growing bacteria and other organisms, 17

Agnatha ('ag-nə-thə) • the class for the jawless fish; also called Marsipobranchii (mär-sə-pō-'bran-kē-ī), 169

alga ('al-gə) [*plural,* **algae** ('al-jē)] • an organism in the kingdom Protista; usually contains chlorophyll, 95-96, 152

alive (ə-'līv) • having life, 2

alternate arrangement • see arrangement, alternate

alternation of generations (ôl-tər-'nā-shən 'əv je-nə-'rā-shənz) • the two-stage sexual reproduction cycle in plants—the sporophyte stage alternates with the gametophyte stage, 121-132

Amoeba (ə-'mē-bə) • the genus for a type protist that moves with pseudopodia, 9

amoeba (ə-'mē-bə) [*plural,* **amoebae** (ə-'mē-bē)] • a protist that moves with pseudopodia, 9, 10, 60, 61, 64

Amphibia (am-'fi-bē-ə) • [Gr., *amphibia,* "living a double life"] the class under the phylum Chordata that contains organisms that live part of their life in water and part on land, 10, 11, 169, 171

amphibian (am-'fi-bē-ən) [Gr., *amphibia,* "living a double life"] an animal in the class Amphibia, 165, 169, 171

anatomy (ə-'na-tə-mē) • the branch of biology that studies the structure of plants and animals; the structural makeup of an organism, 3, 15, 160, 167, 168, 187-188

anatomy (ə-'na-tə-mē), **gross** • the study of the systems and organs of a living thing that are visible to the eye, 187

anatomy, microscopic (ə-'na-tə-mē, mī-krə-'skä-pik) • the study of very small features of living things with the use of a microscope, 187

Ancylostoma duodenale (an-ki-'läs-tə-mə dü-a-də-'na-lē) • a species of hookworm that infects humans, 155

angiosperm ('an-jē-ə-spərm) • a flowering, seeded vascular plant, 87, 89-91, 118, 125, 126, 127, 129-132

animal cell • see cell, animal

Animalia (ä-nə-'māl-yə) • the taxonomic kingdom that contains all the organisms that have animal cells, 6, 8, 10, 11, 45, 148, 151

Annelida (ə-'nel-ə-də) • [L., *annellus,* "little ring"] the phylum for earthworms, leeches, and marine worms, 148, 154, 156

antenna (an-'te-nə) [*plural,* **antennae** (an-'te-nē)] • one of a pair of slender, movable, segmented sensory organs on the head of an arthropod, 152, 156, 158, 160, 161, 162

anther ('an-thər) • in a flower, a small sac that houses microsporangia where male gametes are produced, 130-131

Anthocerotophyta (an-thō-sə-rō-'tä-fə-tə) • the phylum for hornworts, 83, 117, 124

Anura (ə-'nür-ə) • an order in the class Amphibia for the frogs and toads, 171

apical meristem tissue • see tissue, apical meristem

Apicomplexa (ā-pi-kəm-'plek-sə) • a phylum for spore-forming protists, 62

Arachnida (ə-'rak-nəd-ə) • a class in the subphylum Chelicerata; includes spiders, ticks, mites, and scorpions, 161

Araneae (ə-'ra-nē-ē) • the order for spiders in the class Arachnida, 161

Archaea (är-'kē-ə) • in taxonomic classification, the domain that contains the single-celled organisms called archaea; the kingdom is also called Archaebacteria, 6, 9, 10, 42

archaea (är-'kē-ə) • [Gr., *archein,* "the first" or "to rule"] single-celled organisms, some of which live in extreme environments, 6, 8, 9, 10, 42-43, 45, 48-49, 52, 56-57

Archaebacteria (är-kē-bak-'tir-ē-ə) • see Archaea

archaeal prokaryotic cell • see cell, archaeal prokaryotic

Aristotle ('a-rə-stä-təl) • [384-322 B.C.E.] Greek philosopher who thought about how to define life and theorized a moving principle, 3

arrangement, alternate • a leaf arrangement of one leaf per node, 113

arrangement, opposite • a leaf arrangement of two leaves per node, 113

arrangement, whorled • a leaf arrangement of three or more leaves per node, 113

arthropod ('är-thrə-päd) • an animal in the phylum Arthropoda, 151, 158-162

Arthropoda (är-thrə-'pō-də) • [Gr., *arthron,* "joint"; *podos,* "foot"] the phylum under the kingdom Animalia that includes the insects, 10-11, 148, 158-162

Artiodactyla (ärt-ē-ō-'dak-tə-lə) • the order for the even-toed hoofed mammals, 179

ascocarp ('as-kə-kärp) • the fruiting body of an ascomycete, 73, 74

ascomycete (as-kō-'mī-sēt) • a fungus in the phylum Ascomycota, 73-74

Ascomycota (as-kō-mī-'kō-tə) • the phylum that contains yeasts and truffles, 69, 73-74

ascospore ('as-kə-spôr) • a microscopic spore contained in an ascus, 73, 74

chromoplast ('krō-mə-plast) • plastid that contains molecules called carotenoids that give plants, flowers, and fruit their yellow, orange, and red colors, 80

chromosome ('krō-mə-sōm) • a long wound up strand of DNA and protein that is pulled apart for duplication during cell division, 121-124, 138

ciliate ('si-lē-ət) • a protist that moves by using cilia, 60, 61, 65

Ciliophora (sil-ē-'äf-ə-rə) • the phylum for protists that have cilia, 60

cilium (si-lē-əm) [*plural,* **cilia** ('si-lē-ə)] • a small hair-like projection on the body of some protists, 60, 61, 62, 63, 65

Cingulata (sing-yə-'lā-tə) • the taxonomic division for armadillos,

circulatory system ('sər-kyə-lə-tôr-ē 'sis-təm) • the heart and the network of vessels that transport blood throughout the body, 187

class • in taxonomy, a subgroup of a phylum, 10-11, 81, 89, 148, 159, 160, 161, 162, 169, 170, 171, 172, 173, 176

classify ('kla-sə-fī) [**classification** (kla-sə-fə-'kā-shən)] • to arrange into particular groups, 5-6, 10-12, 52, 60- 62, 69, 81-82, 83, 95, 135, 147, 151, 159, 165, 177

clone • a new organism that is genetically the same as the parent, 116, 118, 119

club moss • a small vascular plant in the phylum Lycopodiophyta and the family Lycopodiaceae, 85

Cnidaria (nī-'dar-ē-ə) • [Gr., *cnidos,* "stinging nettle"] the phylum for soft-bodied animals that have cnidocytes in their tentacles, 148, 152-153

cnidocyte ('nī-də-s sīt) • a stinging cell located in the tentacle of a cnidarian, 152-153

cocci ('käk-sē) • [Gr., *kokkus,* "grain, seed, or berry"] sphere-shaped bacteria, 55

coelom ('sē-ləm) • an internal, fluid-filled body cavity in some species of worms, 154, 156

collenchyma cell • see cell, collenchyma

colony ('kä-lə-nē) • a group of individuals that have common characteristics, 54

compass ('kəm-pəs) • a device that uses a magnetic needle to find direction, 16

compound eye • in some arthropods, an eye that is made up of many units, 162

compound leaf • see leaf, compound

compound microscope • see microscope, compound

cone • in conifers, a reproductive structure that produces either pollen or seeds, 87, 127, 128

cone, ovulate ('äv-yə-lāt) • in gymnosperms, the female reproductive structure that produces female gametes (eggs); also called a seed cone, 127-128

cone, pollen • in gymnosperms, the male reproductive structure that produces male gametes (sperm), 127-128

cone, seed • see ovulate cone

conidiophore (kə-'ni-dē-ə-fôr) • structure in which conidia are found, 74, 75

conidium (kə-'ni-dē-əm) [*plural,* **conidia** (kə-'ni-dē-ə) • in some fungi, a spore that is produced asexually, 74, 75

conifer ('kä-nə-fər) • a tree or shrub in the phylum Pinophyta; has needlelike leaves and cones for producing seeds, 87-88, 95, 118, 125, 126-129

contrast ('kän-trast) • amount of difference between light and dark areas, 26

cork cambium • see cambium, cork

cotyledon (kä-tə-'lē-dən) • in a monocot, the part of the seed that develops into the single first leaf; in a dicot, the part of the seed where food is stored for the germinating plant, 90, 91

crepuscular (kri-'pə-skyə-lər) • occurring or active during twilight, 182

crista ('kri-stə) [*plural,* **cristae** ('kri-stē)] • an inwardly projecting fold of the inner membrane of a mitochondrion, 141

Crocodilia (kräk-ə-'dil-ē-ə) • an order in the class Reptilia; contains crocodiles, alligators, caimans, and gavials, 172

crop • in birds, a specialized organ that helps with digestion, 173

cross-striations ('krôs strī-'ā-shənz) • in a muscle cell, overlapping thick and thin protein strands that look like little lines going horizontally across the cell, 146

Crustacea (krə-'stā-shē-ə) • a subphylum of Arthropoda; includes crabs, shrimp, lobsters, and crayfish, 159, 160

cuticle ('kyü-ti-kəl) • in plants, a waxy, waterproof layer covering a leaf, 87, 95

cyanobacterium (sī-ə-nō-bak-'tir-ē-əm) [*plural,* **cyanobacteria** (sī-ə-nō-bak-'tir-ē-ə)] • a single-celled organism in the kingdom Bacteria that uses photosynthesis to make food, 96

cycad ('sī-kəd) • a slow growing, cone-bearing tree in the gymnosperm phylum Cycadophyta, 87, 89, 118, 125

Cycadophyta (sī-kə-'dä-fi-tə) • the taxonomic phylum for cycads, 87, 89, 118, 125

cytoplasm ('sī-tə-pla-zəm) • the water and other molecules that are within the cell but outside the nucleus of the cell, 64, 105, 136, 137, 138

cytoskeleton (sī-tə-'ske-lə-tən) • an array of specialized proteins used for molecular transport, 136, 141-143

cytosol ('sī-tə-säl) • the water-based fluid that fills a cell, 136, 139-140

cytosolic surface (sī-tə-'sä-lik 'sər-fəs) • in the rough endoplasmic reticulum, the outer surface of the membrane that is in contact with the cytosol, 140

dark reaction • see Calvin cycle

Democritus (di-'mä-krə-təs) • [circa 460–370 B.C.E.] Greek philosopher; proposed that all matter is composed of indivisible particles called atoms, 4

dendrite ('den-drīt) • a shorter process (projection) in a neuron; receives signals from surrounding cells, 146-147

deoxyribonucleic acid • see DNA

dermal tissue • see tissue, dermal

dermal tissue system • see tissue system, dermal

Descartes, Rene (dā-'kärt, rə-'nā) • [1596–1650 C.E.] French philosopher; thought about how atoms form molecules; developed the idea of mechanism, 4

dicot ('dī-kät) • an angiosperm that has a seed with two cotyledons and first leaves in a set of two; the term dicot is short for dicotyledon, 89-91

dicotyledon (dī-kä-tə-'lē-dən) • a dicot, 91

Didinium (di-'di-nē-əm) • the genus for a type of ciliate protist, 60

glycerol ether plasma membrane • sse membrane, glycerol ether plasma

glycolysis (glī-'kä-lə-səs) • the conversion of food into energy in living things, 39

Golgi apparatus ('gōl-jē a-pə-'ra-təs) • organelle where proteins are modified, shipped, and stored; also called the Golgi complex, 46, 47, 140, 142

Golgi complex • see Golgi apparatus

GPS • global positioning system; a device that uses satellite signals to determine location, 16

granum ('grā-nəm) [*plural*, **grana** ('grā-nə)] • a stack of thylakoids in a chloroplast, 94

gross anatomy • see anatomy, gross

ground state • the lowest energy state possible for a given atom or molecule, 98

ground tissue • see tissue, ground

Gymnophiona (jim-nə-'fi-ə-nə) • the order for caecilians, 171

gymnosperm ('jim-nə-spərm) • a non-flowering, cone-bearing, seed producing vascular plant, 87-89, 118, 125, 126-129

halophile ('ha-lə-fī-əl) • an organism that lives in a salty environment, 56

hand lens • small handheld magnifier, 16

haploid ('ha-ploid) • a cell or spore that has only one set of chromosomes, 123, 124, 127, 128, 131

Haplorhini (ha-plə-'rī-nē) • a suborder of Primates; includes tarsiers, monkeys, apes, humans, 186

HCNOPS (āch-sē-'nops) **group** • the group of 6 atoms (hydrogen, carbon, nitrogen, oxygen, phosphorus, and sulfur) that make up the majority of biological molecules, 34

Hemichordata (he-mə-kôr-'dā-tə) • a subphylum of Chordata; includes acorn worms and related species, 166, 167

heterotroph ('he-tə-rə-trōf) • an organism that cannot make its own food, 63-65, 134, 151

Hippocrates (hi-'pä-krə-tēz) • [c. 460–c. 377 B.CE.] Greek physician; theorized that life is caused by the ether, 3

homeostasis (hō-mē-ō-'stā-səs) • the state in which an organism maintains a stable internal environment, 188

Homo sapiens (hō-mō-'sā-pē-enz) • the genus and species for humans, 12

Hooke (hək), **Robert** • [1635–1703 C.E.] English scientist; improved the compound microscope; wrote and illustrated his observations in the book *Micrographia*, 24

hookworm • a common roundworm that causes disease in humans, 155

hornwort • nonvascular plant in the phylum Anthocerotophyta, 83-84, 117, 124

horsetail • a vascular plant in the phylum Sphenophyta and the family Equisetaceae, 85

hydrogen ('hī-drə-jən) • smallest atom composed of 1 proton and 1 electron; part of the HCNOPS group, 34, 35, 56

hypha ('hī-fə) [*plural*, **hyphae** ('hī-fē)] • tiny, thread-like structures that make up the mycelium of a fungus, 70, 71-76

Hypocoma (hī-pə-'kō-mə) • the genus for a type of ciliate protist, 60

icosohedral (ī-kō-sə-'hē-drəl) • referring to a solid shape that has 20 faces, 53

immune system (i-'myün 'si-stəm) • the bodily system that protects the body from foreign substances, such as harmful bacteria; the lymphatic system, 15, 145, 187

immunology (i-myə-'nä-lə-jē) • the branch of biology that studies the immune system, 15

incisor (in-'sī-zər) • a chisel-like front tooth used for cutting or tearing food, 176, 183

information molecule • see molecule, information

infraclass • a taxonomic group, 178

insect ('in-sekt) • an animal in the class Insecta; has a segmented body divided into 3 parts, jointed appendages, and an exoskeleton, 70, 108, 129, 130, 131, 132, 134, 161, 162, 173, 178, 183

Insecta (in-'sek-tə) • a class in the subphylum Uniramia; includes all the insects, 159, 162

Insectivora • see Eulipotyphla

integumentary system (in-te-gyə-'men-tə-rē 'sis-təm) • comprised of skin, hair, nails, sweat glands, and oil glands, 188

internode ('in-tər-nōd) • in plants, the part of the stem in between two nodes, 111

Janssen, Hans ('jan-sən, 'hänz) • Dutch spectacle-maker who, along with his son Zacharius, is thought to have invented the compound microscope, 23, 24

Janssen, Zacharias ('jan-sən, za-kə-'rī-əs) • [c. 1580-c. 1638 C.E.] Dutch spectacle-maker; along with his father Hans, is thought to have invented the compound microscope, 23, 24

kelp • protist that groups with others to form a large colony in the ocean, 59, 95

keratin ('ker-ət-ən) • a protein that forms long filaments that become hair, nails, and hooves and forms a protective barrier on the surface of the skin, 145

keratinocyte (kə-'rat-ən-ə-sīt or ker-ə-'tin-ə-sīt) • a skin cell that produces the protein keratin, 145

kinesin (ki-'nē-sən) • a molecular machine that moves cargo around within a cell, 37

kingdom ('king-dəm) • a group in taxonomy that is a subdivision of a domain, 6-12, 45, 52, 59, 60, 68, 69, 79, 81-82, 87, 94, 95, 96, 134, 147, 148, 151

Lagomorpha (lä-gə-'môr-fə) • the order for hares, rabbits, and pikas, 182

Langerhans cell •see cell, Langerhans

lateral bud • see axillary bud

lateral meristem • see meristem, lateral

leaf (*plural*, **leaves**) • the major site of photosynthesis in a plant, 80, 82, 83, 84, 85, 86, 87, 88, 89, 90, 91, 95, 97, 105, 106, 107, 108, 110, 111, 112-113, 117, 118, 119-120, 130

leaf, compound • a leaf that is divided into several leaflets, 113

leaf, scale • see scale leaf

leaf, simple • a leaf with one undivided blade, 113

leaflet • a leaf-like part of a compound leaf

membrane, plasma ('mem-brən, 'plaz-mə) • a thin, soft, greasy film that surrounds a cell and controls what goes into and out of the cell, 43, 44, 45, 46, 49, 136, 137, 141

membrane, semipermeable ('mem-brān, se-mē-'pər-mē-ə-bəl) • a membrane that allows only small molecules to pass through, 110, 137

meristem ('mer-ə-stem) • tissue made of undifferentiated meristematic cells, 109

meristematic cell • see cell, meristematic

meristematic tissue • see tissue, meristematic

meristem, lateral ('mer-ə-stem, 'la-tə-rəl) • meristematic tissue located along the length of stems; produces secondary growth along the sides of a stem to make it thicker; 112

Merostomata (me-rə-'stō-mə-tə) • the class for horseshoe crabs, 160

mesoderm ('me-zə-dərm) • the middle tissue layer of some animals, including flatworms, 154, 155, 156

metabolic (me-tə-'bä-lik) **pathway** • the order followed by chemical reactions in a cell, 39

metabolic reaction (me-tə-'bä-lik rē-'ak-shən) • a chemical reaction in a cell needed to keep the cell alive, 141, 142

metabolism (mə-'ta-bə-li-zəm) • the chemical processes used by the body to stay alive, 38-39

Metamonada (me-tə-mə-'nä-də) • a phylum for flagellates, 61

metamorphosis (me-tə-'môr-fə-səs) • a change from one form to a different form, 162

methane • a gaseous hydrocarbon, 56

methanogen (mə-'than-ə-jen) • a type of archaea that produces methane, 56

metric scale • a system of measurement that uses multiples of 10, 25

microfilament (mī-krō-'fi-lə-mənt) • in a cell, a tiny threadlike structure that contains the protein actin, 141, 143

microscope ('mī-krə-skōp) • an instrument used to magnify objects that can't be seen with unaided eyes, 9, 15, 16, 17, 23- 31, 52, 54, 59, 61, 153, 187

microscope, atomic force (AFM) • a microscope that works by measuring the force between an atom and the tip of the probe, 31

microscope, compound ('mī-krə-skōp, 'käm-pownd) • an instrument that has two or more lenses used together to magnify small objects, 23, 26

microscope, electron ('mī-krə-skōp, i-'lek-trän) • an instrument that uses electron beams to magnify a specimen, 15, 24, 27-28, 52

microscope ('mī-krə-skōp), **light** • an instrument that uses light and the interaction of light with glass lenses to focus and magnify an object, 15, 24, 27-28, 52

microscope, probe ('mī-krə-skōp, 'prōb) • an instrument that uses a probe to "feel" the features on the surface of a sample, 24, 28-31

microscope, scanning electron ('mī-krə-skōp, 'ska-ning i-'lek-trän) **(SEM)** • an instrument in which an electron beam is scanned across a sample surface, a detector picks up the electrons that have been scattered during the scan, and a computer uses this information to create an image, 28

microscope, scanning tunneling (STM) • a type of probe microscope that scans the surface of an object and then projects an image of this surface onto a computer monitor or other screen, 28, 30-31

microscopic (mī-krə-'skä-pik) • describing an object so small that a microscope must be used to see it, 9, 15, 23, 24, 25, 59, 68, 73, 74, 75, 95, 110, 127, 130, 187

microscopic anatomy • see anatomy, microscopic

microscopic scale (mī-krə-'skä-pik skāl) • a standard of measurement for objects that can only be seen with a microscope, 24, 25

microscopy (mī-'kräs-kə-pē) • the use of a microscope for studying small things, 25, 27

Microspora (mī-krä-'spô-rə) • a phylum for spore-forming protists, 62

microsporangiumv (mī-krə-spə-'ran-jē-əm) [*plural,* **microsporangia** (mī-krə-spə-'ran-jē-ə)] • the structure in which a microspore is produced and will develop into a male egg cell during the sexual reproductive cycle of a plant, 122, 123, 128, 130, 131

microspore ('mī-krə-spôr) • a single cell that will develop into a male sperm cell during sexual reproduction in a plant, 122, 123-124, 127, 130

microtubule (mī-krō-'tü-byül) • a long protein tube used by other proteins to move things throughout a cell, 37, 46

mildew ('mil-dü) • a type of fungus, 68

mitochondrion (mī-tə-'kän-drē-ən) [*plural,* **mitochondria** (mī-tə-'kän-drē-ə)] • a small organelle in eukaryotic cells that makes ATP energy molecules, 24, 25, 46, 47, 80, 141, 142

mitosis (mī-'tō-səs) • a process of cell division, 49, 123

molar ('mō-lər) • a tooth used for grinding, 176

mold • a type of fungus in the phylum Zygomycota, 68, 71-72

molecular biology (mə-'le-kyə-lər bī-'ä-lə-jē) • the branch of biology that studies biological molecules, 15, 17-18, 69

molecular machine (mə-'le-kyə-lər ma-'shēn) • a biological molecule that moves other molecules, breaks down unwanted molecules, or makes other molecules, 17, 35, 37, 38, 43, 146

molecule, biological ('mä-li-kyül, bī-ä-'lä-ji-kəl) • a molecule that is part of a living thing, 15, 34-35, 37, 38

molecule, energy ('mä-li-kyül, 'e-nər-jē) • a biological molecule that provides energy for chemical reactions, 35-36, 46, 47, 79, 80, 141

molecule, information • a biological molecule that gives a cell instructions for how to grow and die, 35

molecule, structural ('mä-li-kyül, 'strək-chə-rəl) • a biological molecule that holds a part of a cell together, 35, 36

Mollusca (mə-'ləs-kə) • [L., *molluscus,* "soft"] the phylum for soft-bodied animals with internal or external shells; includes clams, snails, squids, and octopuses, 148, 156-158

oral surface • the lower surface of an echinoderm that contains the mouth and the tube feet, 162

order • in taxonomy, a subgroup that is a division of a class, 10, 11, 81, 148, 159, 161, 171, 172, 173, 177, 178-186

organ (ôr-gən) • a group of tissues that together perform a particular function, 43, 109, 112, 134, 143, 145, 151, 156, 157, 158, 159, 168, 170, 171, 172, 173, 176, 177, 178, 187-188

organelle (ôr-gə-'nel) • a structure that performs a specific function inside a eukaryotic cell, 43, 45-47, 79-80, 94, 105, 134, 136, 137, 138-143, 152

organic (ôr-'ga-nik) compound • a carbon-containing molecule, 134

organ system ('ôr-gən 'sis-təm) • a group of cells, tissues, and organs that perform a particular function in an organism, 187-188

osmosis (äz-'mō-səs) • a physical process in which molecules pass through a membrane, 110

Osteichthyes (ä-stē-'ik-thē-ēz) • a class in the subphylum Vertebrata; comprised of the bony fishes; also called Pisces ('pī-sēz), 170

osteoblast ('ä-stē-ə-blast) • a bone-making cell, 144

osteoclast ('ä-stē-ə-klast) • a cell involved in breaking down bone tissue during growth and healing, 144

osteocyte (ä-stē-ə-'sīt) • a bone cell, 144

ovary ('ō-və-rē) • the part of a flowering plant where the seeds form, 131

ovulate cone • see cone, ovulate

ovule ('äv-yül) • in the sexual reproductive cycle of a plant; contains the megasporangium in which the female gamete (egg) develops and when fertilized becomes a seed, 128, 131

oxygen ('äk-si-jən) • an atom in the HCNOPS group and in the atmosphere; given off by photosynthesis, 34, 35, 38, 56, 96, 144, 154, 156, 158, 168, 170, 178, 187, 188

Panthera (pan-'thē-rə) • a genus in the family Felidae, 12

Panthera leo (pan-'thē-rə 'lē-ō) • the taxonomic name (genus and species) for a lion, 12

Panthera tigris (pan-'thē-rə 'tī-gris) • the taxonomic name (genus and species) for a tiger, 12

Paramecium (per-ə-'mē-shē-əm) • the genus for a type of protist that has cilia, 60

paramecium (per-ə-'mē-shē-əm) [plural, paramecia (per-ə-'mē-shē-ə) • a type of protist that has cilia, 9, 60, 63, 64, 65

parasite ('per-ə-sīt) • an organism that feeds on living things, 62, 69

parenchyma cell • see cell, parenchyma

pathogen ('pa-thə-jən) • a disease-causing microorganism, 112, 145

pedipalp ('pe-də-palp) • for animals in the subphylum Chelicerata, one of a pair of appendages used to grab prey, 160, 161

pellicle ('pe-li-kəl) • a thin outer covering on some protists, 61

periderm ('per-ə-dərm) • in woody plants, a protective tissue that replaces the epidermis in older regions of the stem and roots, 108

Perissodactyla (pə-ris-ə-'dak-tə-lə) • the order for odd-toed ungulate mammals, 180

permeable membrane • see membrane, permeable

peroxisome (pə-'räk-sə-sōm) • an organelle that contains enzymes that are involved in a number of metabolic reactions in a cell, 47, 80, 141, 142

petal ('pe-təl) • one of a group of modified leaves that surround the reproductive organs of a flower; typically brightly colored, 130

petiole ('pe-tē-ōl) • a stalk that attaches the blade of a leaf to the stem at a node, 113

phagocyte ('fa-gə-sīt) • [Gr., phagein, "to eat"; Gr., kytos, "receptacle or container"] a cell that eats other material, 64

phagocytosis (fa-gə-sə-'tō-səs) • the process of eating food by surrounding it, 64

pharyngeal (fer-ən-'jē-əl or fə-'rin-jē--əl) pouch • in fishes and amphibians, tissues in the throat region that develop into gills, 165, 167

pharynx ('fer-inks) • for vertebrate animals, the cavity behind the mouth opening where oxygen and food enter, 168

philosophical (fi-lə-'sä-fi-kəl) map • a particular way to interpret the world around us, 5

phloem ('flō-em) • the vascular tissue that transports sugars made by photosynthesis downward through the plant all the way to the roots, providing food for the plant, 82, 84, 90, 108, 111-112

phosphate ('fäs-fāt) bond • the bond between two phosphate groups in a biological molecule; when this type of bond is broken, energy is released, 35-36

phosphorus ('fäs-fə-rəs) • an atom in the HCNOPS group, 34, 35

photon ('fō-tän) • a packet of energy that travels through space like a wave, 98, 100, 101

photosynthesis (fō-tō-'sin-thə-səs) • [Gr., photos, light; Gr., synthesis, to make] the process by which plants convert energy from the Sun into food, 39, 46, 62, 63, 71, 80, 87, 94-102, 105, 106, 108, 111, 112, 152

photosystem ('fō-tō-si-stəm) • a complex group of proteins and other molecules working together to make sugar, 100-102

photosystem I ('fō-tō-si-stəm 'wən) • the second of two functions in the light-dependent reactions for photosynthesis, 100-102

photosystem II ('fō-tō-si-stəm 'tü) • the first of two functions in the light-dependent reactions for photosynthesis, 100-102

phylum ('fī-ləm) [plural, phyla ('fī-lə) • in taxonomy, a division of a kingdom, 10, 11, 60-62, 69, 70, 71-76, 81-91, 117, 118, 124, 147-148, 151-163, 165-169

physiology (fi-zē-'ä-lə-jē) • the branch of biology that studies how plant and animal bodies function, 15

pigment ('pig-mənt) • a molecule that provides color to living things, 80, 98-99

pseudocoelom (sü-də-'sē-ləm) • in roundworms, the internal cavity only partially layered with mesoderm tissue; a "false coelom," 155

pseudopod ('sü-də-päd) [plural, pseudopodia (sü-də-'pō-dē-ə)] • [Gr., pseudo, "false"; Gr., podia, "feet"] in an amoeba, an extension of the edge of the membrane, 61, 62 64

Psilophyta (sil-'ä-fə-tə) • the phylum for whisk ferns, 86, 117, 124

Pterobranchia (te-rə-'bran-kē-ə) • a taxonomic group of hemichordates; the animals are very small and live in large colonies, 167

radar equipment ('rā-där i-'kwip-mənt) • instruments that use radio waves to collect data, 16

radial body symmetry • see symmetry, radial body

radula ('ra-jə-lə) • in some mollusks, a tongue-like structure that is covered in rows of small teeth, 157

reaction center complex • a specific section of photosystem I where electrons are excited to a high energy state, 100-101

red blood cell • see cell, red blood

reductionism (ri-'dək-shə-ni-zəm) • the belief that because life can be explained by the laws of chemistry and physics (materialism) you can completely understand something by studying its parts, 4, 5

reflect (ri-'flekt) • in physics, for a light wave to bounce off a material, 97, 98, 99

remotely operated vehicle (ROV) • a device that can be operated from a distance, 19

reproduction, asexual (rē-prə-'dək-shən, ā-'sek-shə-wəl) • a type of reproduction in which an organism does not combine its genetic material with the genetic material of another organism, 70, 72, 74, 75, 83, 116-121

reproduction, sexual (rē-prə-'dək-shən, 'sek-shə-wəl) • a type of reproduction in which genetic material is shared between two organisms, 70-75, 83, 116, 121-132, 134

reproductive system (rē-prə-'dək-tiv 'sis-təm) • produces sperm (male) and eggs (female) used to create new organisms, 188

reptile ('rep-tī-əl) • an animal in the class Reptilia; has a vertebral column, dry scaly skin, and developed lungs,

Reptilia (rep-'ti-lē-ə) • a class in the subphylum Vertebrata; includes snakes, lizards, crocodiles, turtles, and tortoises, 172

resolution (re-zə-'lü-shən) • the ability of a microscope to separate fine details of a sample, 27, 28

resolve (ri-'zälv) • to make clear, 24, 26, 27

respiratory system ('re-spə-rə-tôr-ē 'sis təm) • the lungs are the main organs of this system; oxygenates blood and eliminates carbon dioxide, 188

rhinovirus (rī-nō-'vī-rəs) • a type of virus that causes the common cold, 53

rhizoid ('rī-zoid) • a small hairlike structure on a nonvascular plant that digs into the soil to keep the plant in place, 83, 84

rhizome ('rī-zōm) • a creeping horizontal stem that usually grows underground and can produce roots and shoots from nodes, 85, 86, 117, 118, 119, 121

ribonucleic acid • see RNA

ribose ('rī-bōs) sugar • part of a nucleotide, 38

ribosome ('rī-bə-sōm) • a molecule involved in protein synthesis, 47, 140, 142

ribulose 1,5-bisphosphate ('rib-yə-lōs wən-fiv-bis-'fäs-fāt) • a molecule involved in the Calvin cycle, 102

RNA (ribonucleic acid) (rī-bō-nü-'klē-ik 'a-səd) • a special type of biological molecule used by the cell to store and transfer information, 17, 18, 28, 38, 52, 138

Rodentia (rō-'den-chē-ə) • (L., rodere, "to gnaw") the order that includes squirrels, mice, and rats, 183

root • the part of a plant that grows in soil and absorbs water and nutrients, 71, 90, 105-110, 112, 118, 120-121

root, fleshy • a root that has a thickened taproot, such as a carrot or beet, 121

root, tuberous ('tü-bə-rəs) • a root that has thickened areas on secondary roots that branch out laterally from the main root, 121

rough endoplasmic reticulum • see endoplasmic reticulum, rough

roundworm • a worm in the phylum Nematoda that has a rounded body and a pseudocoelom between the endoderm and mesoderm, 148, 155

ROV • see remotely operated vehicle

rumen ('rü-mən) • in some herbivore mammals, a compartment of the stomach, 176

runner • see stolon

saprophyte ('sa-prə-fīt) • an organism that uses dead or decaying matter for food, 69

scale leaf • a small modified leaf structure, 119

scanning electron microscope (SEM) • see microscope, scanning electron

scanning tunneling microscope (STM) • see microscope, scanning tunneling

sclerenchyma cell • see cell, sclerenchyma

Scorpiones (skôr-pē-'än-əs) • the order for scorpions; in the class Arachnida, 161

scuba ('skoo-bə) equipment • [self-contained underwater breathing apparatus] a portable apparatus that contains gas for a diver to breathe while under water, 19

seaweed • a large colony of protists that lives in the ocean, 59, 95

secondary growth • the process by which plant stems increase in diameter; produced by lateral meristems, 112

sediment corer ('se-də-mənt 'kôr-ər) • a tool used to drill a cylindrical sample from the ocean floor, 19

seed cone • see cone, ovulate

seeded vascular plant • see, vascular plant, seeded

SEM • see microscope, scanning electron

semipermeable membrane • see membrane, semipermeable

vascular ('va-skyə-lər) **plant, non-seeded** • a plant without xylem and phloem; does not use seeds to reproduce, 82, 84, 85-86, 117-118, 124-125

vascular ('va-skyə-lər) **plant, seeded** • a plant with xylem and phloem; uses seeds to reproduce, 82, 84, 87-91, 118

vascular tissue • see tissue, vascular

vascular tissue system • see tissue system, vascular

vegetative propagation ('ve-jə-tā-tiv prä-pə-'gā-shən) • a type of asexual reproduction that occurs when part of a plant breaks off and develops into a new plant, 83, 117-121, 125, 127

vein • in plants, the part of a leaf that contains the vascular tissues and also provides support for the leaf, 90, 106, 113

vertebra ('vər-tə-brə) [*plural,* **vertebrae** ('vər-tə-brā)] • one of the bones in the spinal column of an animal, 147

vertebral column ('vər-tə-brəl 'kä-ləm) • a series of interconnected bones that run down the back of animals in the phylum Chordata, 134, 165, 166, 172

Vertebrata (vər-tə-'brät-ə) • the subphylum of Chordata that includes animals that have a vertebral column, 166, 169-173, 176-188

vertebrate ('vər-tə-brət) • an animal that has a spinal column surrounding a spinal cord, 147, 166, 169-173, 176-188

vesicle ('ve-si-kəl) • a membrane-bound spherical sac involved in transporting and processing proteins in a cell, 140-141

vibrio ('vi-brē-ō) • a type of bacteria that is curved, 56

virus (vī-rəs) • a tiny sack containing DNA or RNA that is surrounded by a tough outer coat made of protein and has many characteristics of a cell, 52-53, 145, 188

vitalism ('vī-tə-li-zəm) • the idea that an unknown force (vital spirit) is responsible for making creatures be alive, 5

vital spirit ('vī-təl 'spir-ət) • the moving principle of life as defined by Galen; the idea that led to vitalism, 3

whisk fern • a vascular plant in the phylum Psilophyta, 86, 117, 124

white blood cell • see cell, white blood

whorled arrangement • see arrangement, whorled

Woese ('wōz), **Carl** • [1928–2012 C.E.] American microbiologist who developed a system of taxonomy that was introduced in 1990, 6

Xenarthra (zə-'när-thrə) • [Gr., *xenos,* "foreign, or alien," *arthron,* "joint"] suborder of the order Eulipotyphla; includes sloths, anteaters, armadillos; formerly Edentata (ē-den-'tä-tə), 184

xylem ('zī-ləm) • the vascular tissue that transports water and nutrients from the roots throughout the plant body, 82, 84, 90, 106, 108, 110-112

yeast • a fungus in the phylum Ascomycota, 68, 70, 74

zoology (zō-'ä-lə-jē) • the branch of biology that studies animals, 15

zygomycete (zī-gō-'mī-sēt) • a member of the phylum Zygomycota, 71-72

Zygomycota (zī-gō-mī-'kō-tə) • the phylum for molds, 69, 71-72

zygosporangium (zī-gō-spə-'ran-jē-əm) [*plural,* **zygosporangia** (zī-gō-spə-'ran-jē-ə) • a spore from which a new mycelium can grow, 72

zygote ('zī-gōt) • a cell that contains genetic material from both the sperm cell and the egg cell; a fertilized egg cell, 123-124, 128, 131

Pronunciation Key

a	add	k	cool	ü	sue
ā	race	l	love	v	vase
ä	palm	m	move	w	way
â(r)	air	n	nice	y	yarn
b	bat	ng	sing	z	zebra
ch	check	o	odd	ə	a in above
d	dog	ō	open		e in sicken
e	end	ô	jaw		i in possible
ē	tree	oi	oil		o in melon
f	fit	oo	pool		u in circus
g	go	p	pit		
h	hope	r	run		
i	it	s	sea		
ī	ice	sh	sure		
j	joy	t	take		
		u	up		

More REAL SCIENCE-4-KIDS Books
by Rebecca W. Keller, PhD

Building Blocks Series yearlong study program — each Student Textbook has accompanying Laboratory Notebook, Teacher's Manual, Lesson Plan, Study Notebook, Quizzes, and Graphics Package

Exploring Science Book K (Activity Book)
Exploring Science Book 1
Exploring Science Book 2
Exploring Science Book 3
Exploring Science Book 4
Exploring Science Book 5
Exploring Science Book 6
Exploring Science Book 7
Exploring Science Book 8

Focus On Series unit study program — each title has a Student Textbook with accompanying Laboratory Notebook, Teacher's Manual, Lesson Plan, Study Notebook, Quizzes, and Graphics Package

Focus On Elementary Chemistry
Focus On Elementary Biology
Focus On Elementary Physics
Focus On Elementary Geology
Focus On Elementary Astronomy

Focus On Middle School Chemistry
Focus On Middle School Biology
Focus On Middle School Physics
Focus On Middle School Geology
Focus On Middle School Astronomy

Focus On High School Chemistry

Super Simple Science Experiments

21 Super Simple Chemistry Experiments
21 Super Simple Biology Experiments
21 Super Simple Physics Experiments
21 Super Simple Geology Experiments
21 Super Simple Astronomy Experiments
101 Super Simple Science Experiments

Note: A few titles may still be in production.

Gravitas Publications Inc.
www.gravitaspublications.com
www.realscience4kids.com